Laboratory Atlas
_of_ Anatomy
and Physiology

THIRD EDITION

Douglas J. Eder, PhD
Southern Illinois University—Edwardsville

Shari Lewis Kaminsky, DPM

John W. Bertram

McGraw Hill

Boston   Burr Ridge, IL   Dubuque, IA   Madison, WI   New York   San Francisco   St. Louis
Bangkok   Bogotá   Caracas   Lisbon   London   Madrid
Mexico City   Milan   New Delhi   Seoul   Singapore   Sydney   Taipei   Toronto

McGraw-Hill Higher Education

A Division of The **McGraw-Hill** Companies

LABORATORY ATLAS OF ANATOMY AND PHYSIOLOGY
THIRD EDITION

1 2 3 4 5 6 7 8 9 0 QPD/QPD 0 9 8 7 6 5 4 3 2 1

ISBN 0–07–290755–X

Vice president and editor-in-chief: *Kevin T. Kane*
Publisher: *Michael D. Lange*
Sponsoring editor: *Kristine Tibbetts*
Developmental editor: *Patricia Hesse*
Marketing managers: *Michelle Watnick/Heather K. Wagner*
Senior project manager: *Marilyn M. Sulzer*
Production supervisor: *Sandy Ludovissy*
Designer: *K. Wayne Harms*
Cover designer: *Rokusek Design*
Photo research coordinator: *John C. Leland*
Compositor: *Interactive Composition Corporation*
Typeface: *10/12 Goudy*
Printer: *Quebecor Printing Book Group/Dubuque, IA*

Photo credits:
Chapter 1: All photos by Ed Reschke except figures 1–89 and 1–90
Chapters 2 and 4: Douglas Eder, Shari Lewis Kaminsky, and John Bertram
All chapter opener photos by Ed Reschke

Library of Congress Cataloging-in-Publication Data

Eder, Douglas J.
    Laboratory atlas of anatomy and physiology/ Douglas J. Eder, Shari Lewis Kaminsky,
John W. Bertram.— 3rd ed.
        p.   cm.
    Includes index.
    ISBN 0–07–290755–X
    1. Musculoskeletal system—Atlases.   I. Kaminsky, Shari Lewis.   II. Bertram, John W.
III. Title.

WM100.E24 2001
611'.7'0222—dc21

00–038014
CIP

CONTENTS

Dedication, iv
Acknowledgments, iv

CHAPTER 1

Histology, 1

CHAPTER 2

Human Skeletal Anatomy, 45

CHAPTER 3

Human Muscular Anatomy, 65

CHAPTER 4

Dissections, 89

CHAPTER 5

Reference Tables, 135

Index, 163

DEDICATION

To Suzanne, Nicholas & Daniel, and Mary Ellen

ACKNOWLEDGMENTS

We obtained all animal material pictured from Nasco Company, Ft. Atkinson, Wisconsin. Cat cadavers were skinned at the factory and packed in a non-formaldehyde preservative. At our request, Nasco personnel selected particularly well-injected cadavers for us; we thank them for this service. We would also like to thank our colleague, Dave Bolt of West Hills College, who reviewed our animal dissections.

Sarah M. Bales
Moraine Valley Community College

Timothy A. Ballard
UNC-Wilmington

R. P. Benard
American International College

Dave Bolt
West Hills College

Robert A. Boomsma
Trinity Christian College

John R. Capeheart
University of Houston-Downtown

Clementine A. de Angelis
Tarrant County College

Dalia Giedrimiene
Saint Joseph College

Donald Glassman
Des Moines Area Community College

Safawo Gullo
Abraham Baldwin College

Albert Isaak
Bob Jones University

George Karleskint, Jr.
St. Louis Community College at Meramec

Carol A. Kravits
Butler County Community College

Roger L. Lane
Kent State University-Ashtabula Campus

Marian G. Langer
St. Francis College of PA

Jerri K. Lindsey
Tarrant County Junior College-NE Campus

Wayne Meyer
Austin College

Pamela J. Reed
Delaware Valley College

John M. Ripper
Butler County Community College

Janice Yoder Smith
Tarrant County Junior College, Northwest

Janet Steele
University of Nebraska at Kearney

Jeffrey A. Taylor
North Central State College

Mary L. Truex
Indiana University South Bend

Timothy T. Turner
Allegany College of Maryland

Mary T. Weis
Collin County Community College District

Jeanne M. Workman
Duquesne University

Contributions by
Bruce D. Wingerd
San Diego State University

Histology

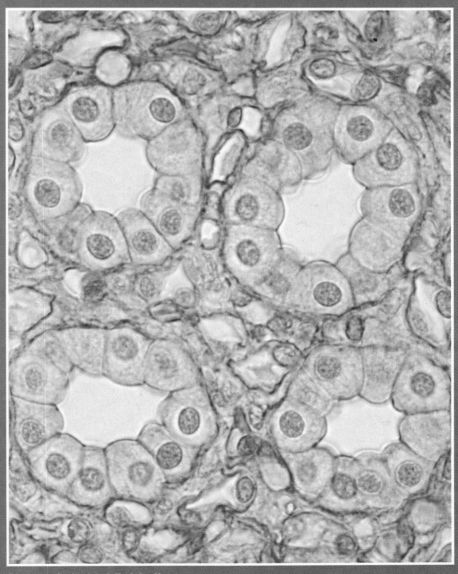

Simple Cuboidal Epithelium

Figure 1-1
Interphase   Nuclear membrane intact with chromatin visible. (×250)

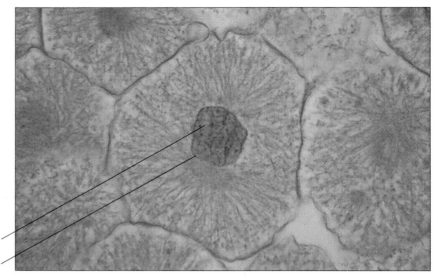

Chromatin

Nuclear membrane

Figure 1-2
Prophase   Duplicated chromosomes condensed into visible strands; nuclear membrane absent. (×250)

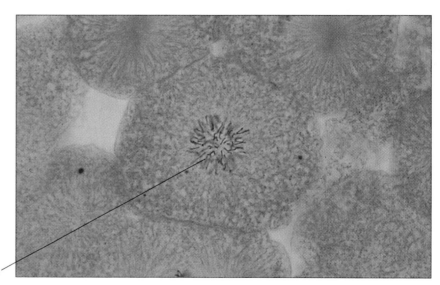

Chromosome

Figure 1-3
Metaphase   Darkly stained chromosomes positioned by microtubular framework to align at cell equator. Spindle fibers and aster visible. (×250)

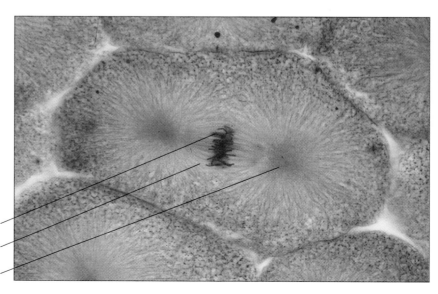

Chromosomes

Spindle fibers

Aster

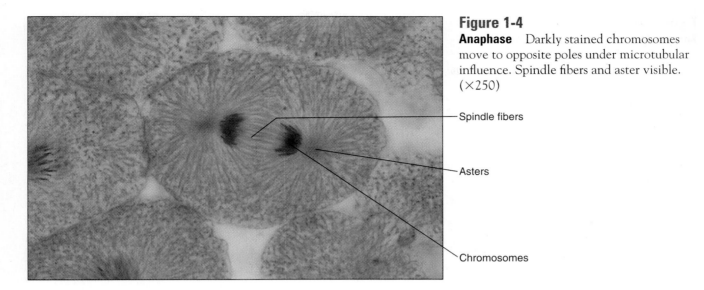

Figure 1-4
Anaphase Darkly stained chromosomes move to opposite poles under microtubular influence. Spindle fibers and aster visible. ($\times$250)

Spindle fibers

Asters

Chromosomes

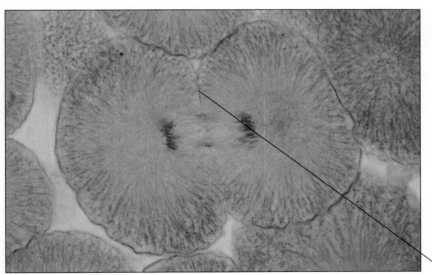

Figure 1-5
Telophase Separated chromosomes lose microtubular attachments. Belt of actinomyocin forms at equator, assists in formation of new cell membranes and cytokinesis. Cleavage furrow forms two daughter cells. ($\times$250)

Cleavage furrow at equator

Figure 1-6
Simple Squamous Epithelium   Single layer of flat cells covering a surface. From human omentum. (×250)

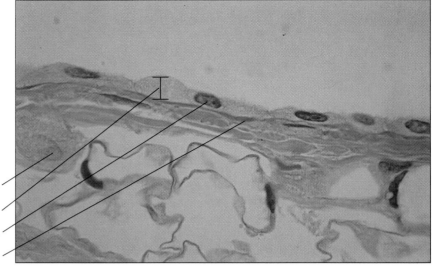

Fixed macrophage

Simple squamous epithelium

Nucleus

Basement membrane

Figure 1-7
Simple Squamous Epithelium   Surface view of flattened cells. Human mesothelium. (×250)

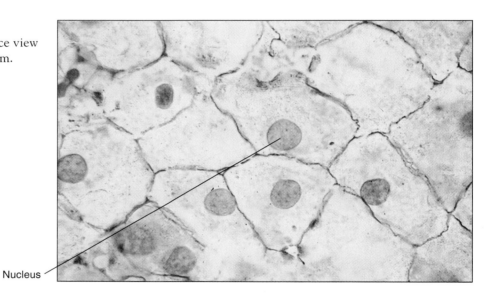

Nucleus

Figure 1-8
Simple Cuboidal Epithelium   Although not strictly cube shaped, cuboidal cells are roughly equidimensional in length, width, and depth. Single layer of cells lining surface of kidney tubules. Cross section. (×250)

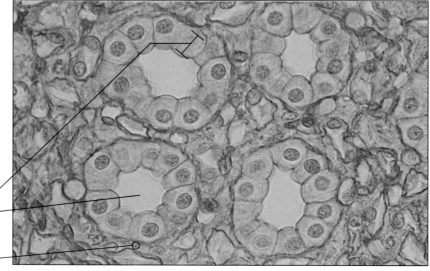

Simple cuboidal epithelium

Lumen of tubule

Basement membrane (circled)

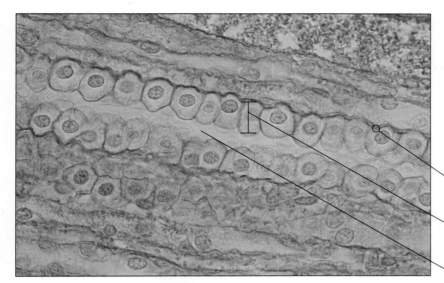

Figure 1-9
Simple Cuboidal Epithelium   Longitudinal section of kidney tubule. (×250)

Basement membrane

Simple cuboidal epithelium

Lumen of tubule

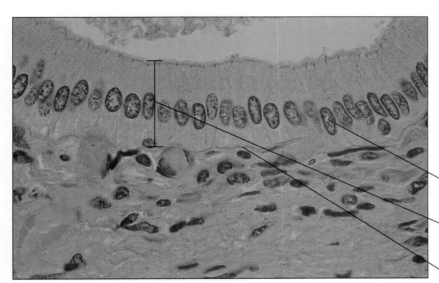

Figure 1-10
Simple Columnar Epithelium   Cellular height is much greater than width or length. Nuclei generally appear in a row. From pancreatic duct. (×250)

Nucleus

Simple columnar epithelium

Basement membrane

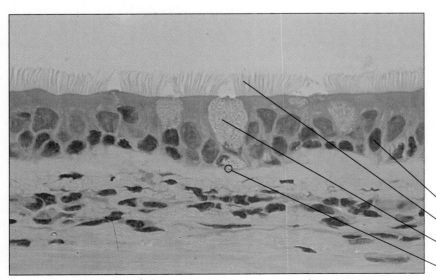

Figure 1-11
Pseudostratified Ciliated Columnar Epithelium   Nuclei appear to lie in two rows, but in fact all cells in single layer are in contact with basement membrane. Section shows well-defined cilia, three goblet cells, basement membrane, underlying connective tissue. From monkey trachea. (×100)

Nucleus

Cilia

Goblet cell

Basement membrane

Figure 1-12
Pseudostratified Ciliated Columnar Epithelium   Section shows cilia, multiple layers of nuclei, basement membrane, underlying connective tissue. From human trachea. (×250)

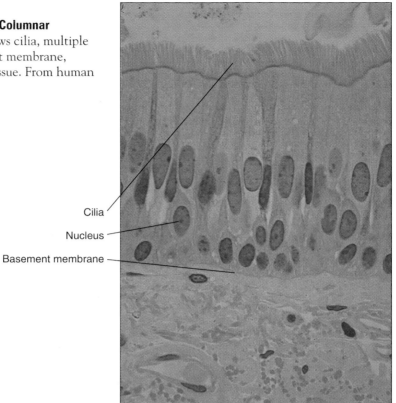

Cilia

Nucleus

Basement membrane

Figure 1-13
Stratified Squamous Epithelium   Flattened cells at surface change to less flattened morphology in deeper layers. Oral cavity of rabbit. (×100)

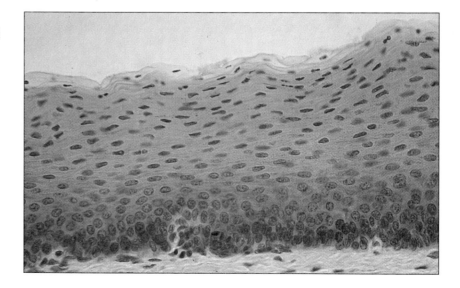

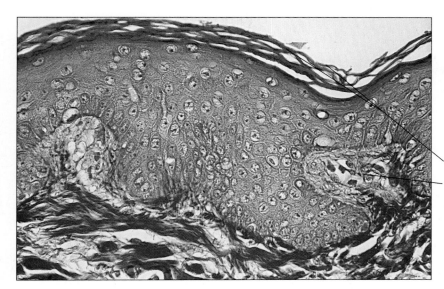

Figure 1-14
Stratified Squamous Epithelium   Flattened, keratinized cells at surface show variations in form in deeper layers. From human skin. (×100)

Keratinized cells

Papilla

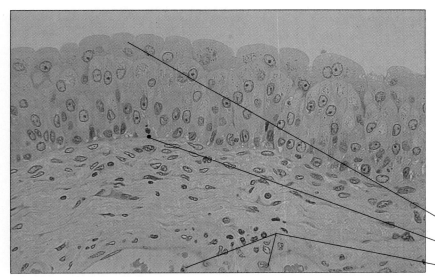

Figure 1-15
Transitional Epithelium from Urinary Bladder   Umbrella cells stretch and flatten as bladder fills. Basement membrane separates epithelium from underlying connective tissue containing blood vessels. (×250)

Umbrella cell

Basement membrane

Blood vessel lumen

Figure 1-16
Wall of Elastic Artery   Extracellular elastic fibers running parallel in a plane. Structure permits tissue elasticity and recoil. From aorta. (×100)

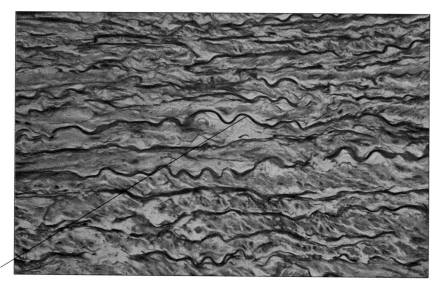

Elastic fiber

Figure 1-17
Reticular Connective Tissue   Mesh of reticular fibers appears as dark lines; provides scaffold for cellular organization of this lymph node. (×250)

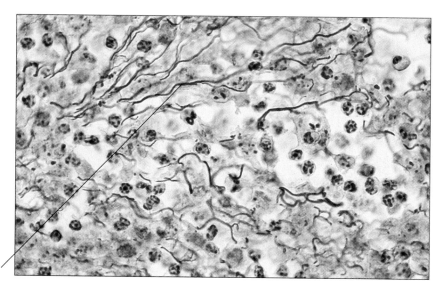

Reticular fiber

Figure 1-18
Loose (Areolar) Connective Tissue   Pink bands of collagen fibers run in all directions through intercellular spaces of subcutaneous tissue, permit flexible resistance to mechanical stress. (×100)

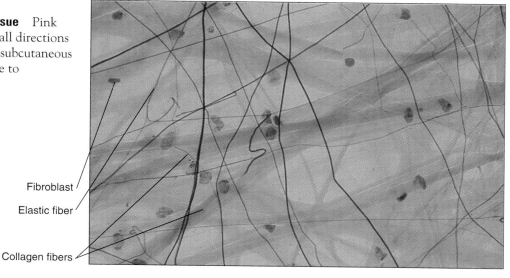

Fibroblast

Elastic fiber

Collagen fibers

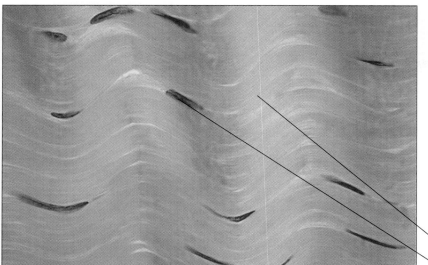

Figure 1-19
Dense Regular Connective Tissue   Bands of collagen fibers running in regular, parallel rows resist mechanical stress mainly along course of fibers. Monkey tendon. (×250)

Collagen fibers
Nucleus of fibroblast

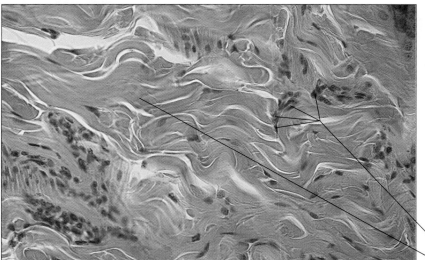

Figure 1-20
Dense Irregular Connective Tissue   Bands of collagen running in irregular rows give multidirectional tensile strength. Collagen-secreting fibroblasts appear throughout. (×100)

Nuclei of fibroblasts
Collagen fibers

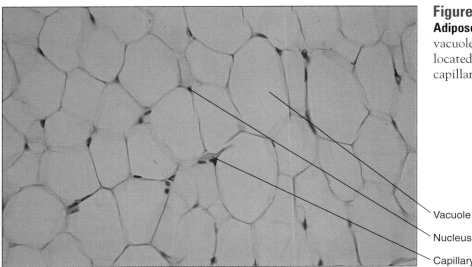

Figure 1-21
Adipose Tissue   Large, empty, polyhedral vacuoles dominate small, eccentrically located cell nuclei of adipocytes. Fine capillaries run through tissue. (×100)

Vacuole
Nucleus
Capillary

Figure 1-22

Fibrocartilage   Cell nests of chondrocytes in territorial matrix surrounded by coarse extracellular fibers. (×250)

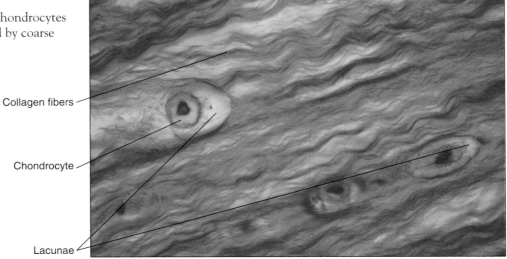

Collagen fibers

Chondrocyte

Lacunae

Figure 1-23

Hyaline Cartilage   During interstitial growth, cartilage cells often form small clusters and move apart as they secrete extracellular matrix. (×100)

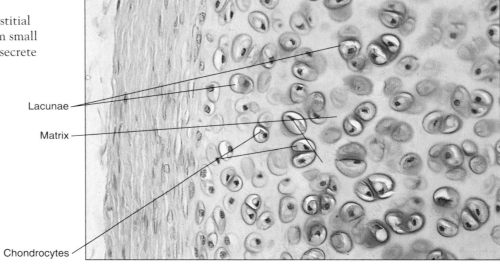

Lacunae

Matrix

Chondrocytes

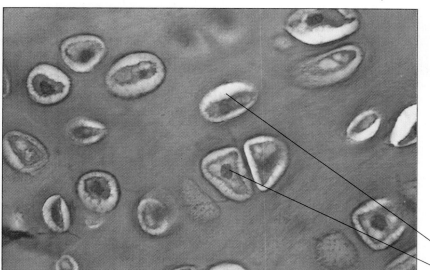

Figure 1-24
Hyaline Cartilage   Artifactual vacuolation forms characteristic lacunae around chondrocyte cell bodies. From trachea. (×250)

Lacuna

Chondrocyte

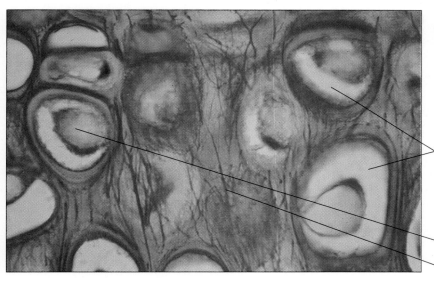

Figure 1-25
Elastic Cartilage   Extracellular matrix contains elastic fibers that confer elastic recoil to this tissue. (×250)

Lacunae

Chondrocyte

Elastic fiber

Figure 1-26
Skin   Thick, keratinized, multilayered
stratum corneum rests atop grainy **stratum
granulosum** (stratum lucidum not clearly
evident). **Stratum spinosum,** composed of
irregularly shaped cells with indistinct
nuclei, lies atop single, clearly nucleated
stratum basale. Human palm. (×100)

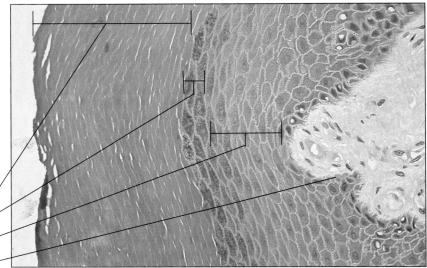

Stratum corneum

Stratum granulosum

Stratum spinosum

Stratum basale

Figure 1-27
Skin   Squamous epidermis with cornified
layers overlying darkly stained stratum
basale and connective tissue of underlying
dermis. Single papilla visible. Human
scalp. (×100)

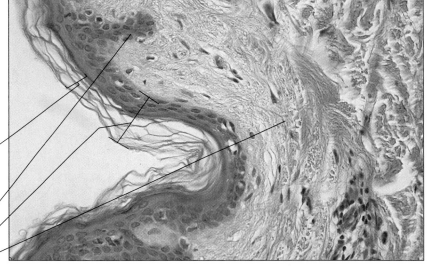

Cornified layer

Papilla

Epidermis

Dermis

Figure 1-28
Meissner's Corpuscle in Dermis   Elongated
oval body located in dermis just below
stratum basale is thought to be responsible
for part of fine touch reception. (×100)

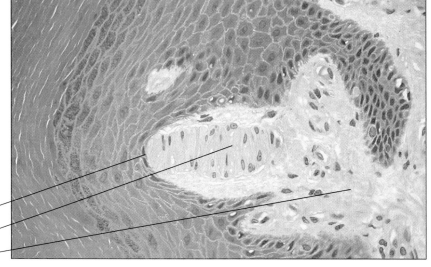

Stratum basale (epidermis)

Meissner's corpuscle

Dermis

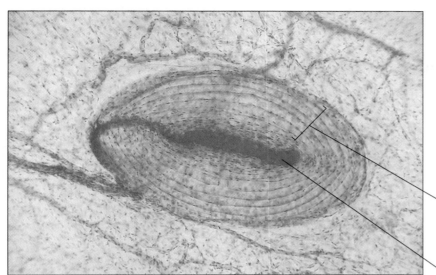

Figure 1-29
Pacinian Corpuscle Encapsulated nerve ending found deep in dermis and throughout interior of body detects pressure. (×25)

Capsule

Free nerve ending

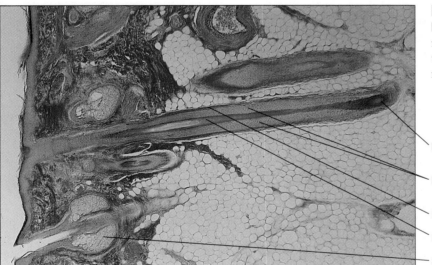

Figure 1-30
Human Scalp with Hair Follicle Follicle root, with sheath embedded in pale adipose tissue, has sebaceous glands surrounding it near surface. (×10)

Hair papilla

Hair follicle

Root sheath

Hair root

Sebaceous gland

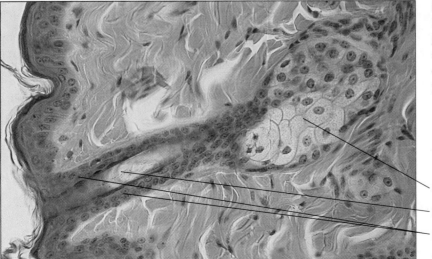

Figure 1-31
Detail of Sebaceous Gland Nucleated germinative cells at base of gland mature and accumulate lipid. At duct, they degenerate and lyse to release their oily product, sebum. (×100)

Sebaceous gland

Hair root

Hair follicle

Figure 1-32

Compact Bone   Center of "tree ring" structure, Haversian canal contains blood vessel. Osteocytes imprisoned in small, dark lacunae surrounding central Haversian canal receive nutrition and communicate via canaliculi, or little canals. Human. (×50)

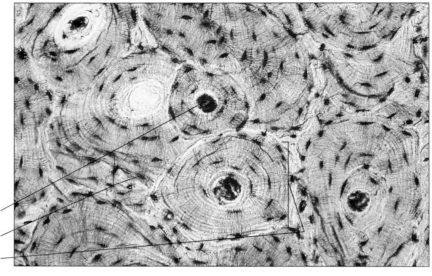

Haversian canal

Lacunae

Haversian system (osteon)

Figure 1-33

Detail of Compact Bone   Haversian system evident.

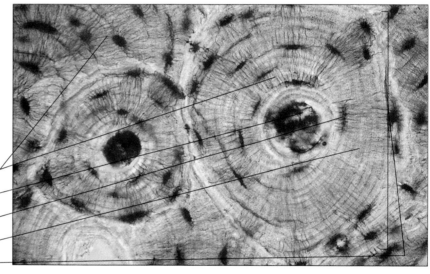

Canaliculus

Haversian canal

Osteocyte in lacuna

Lamella

Haversian system (osteon)

Figure 1-34

Cancellous (Spongy) Bone   Osteoblasts on spongy bone are engaged in secretion of new bony matrix. (×100)

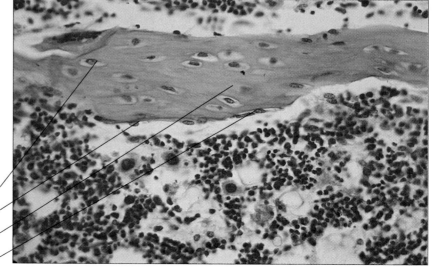

Osteocyte

Resting osteoblast

Spongy bone

Osteoblast

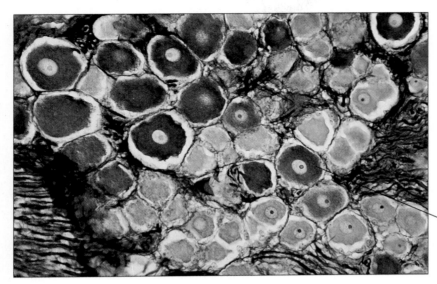

Figure 1-47
Dorsal Root Ganglion   Sensory signals representing pain, temperature, pressure, muscle tension, joint position, and others depend on these cells. Their dendrites collect sensory information throughout the body and axons route it into the spinal cord. (×100)

— Cell body of neuron

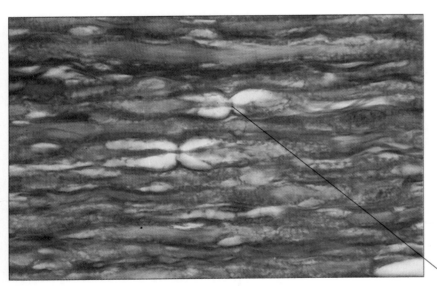

Figure 1-48
Nerve Fibers (Longitudinal Section)   Clear areas show dimpling characteristic of nodes of Ranvier. (×250)

Node of Ranvier

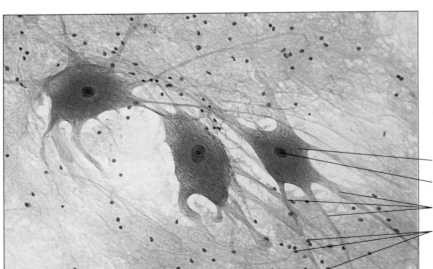

Figure 1-49
Motor Neurons of the Spinal Cord
Integrated command information from the brain and sensory signals enter these cells, whose efferent activity controls muscular contraction. Numerous synapses occur on dendrites and cell body (soma). (×50)

Cell body of neuron

Nucleus

Neuronal processes

Neuroglia

Figure 1-50
Myelinated Nerve Fibers (Cross Section)
Central core stains dark; insulating myelin appears white. (×250)

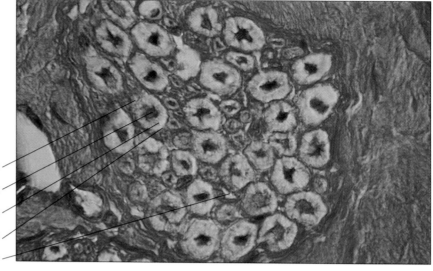

Core of nerve
Axon
Myelin sheath
Neurilemma
Capillary

Figure 1-51
Spinal Cord, Lumbar Region (Cross Section)
Top is dorsal, bottom is ventral. Light central dot is central canal. Darkly staining H-shaped region is grey matter of cell bodies; surrounding lighter material is composed of myelinated axons. Ventral horns of gray matter contain motor neurons; dorsal horns contain cell bodies of sensory pathways. (×4)

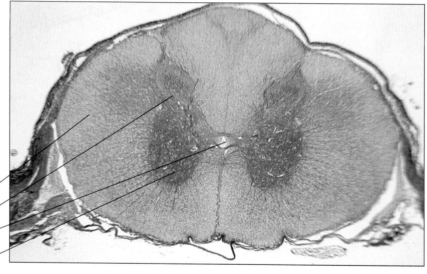

White matter
Dorsal horn
Central canal
Ventral horn

Figure 1-52
Retina   Layered structure evident. Dark line of cells near top is pigment epithelium. Broad striped region represents photoreceptors (rods and cones), whose nuclei stain heavily immediately beneath. Below receptor nuclei lie synaptic region and a layer of nuclei belonging to bipolar cells. Bipolar cell output synapses onto ganglion cells, only a few of which appear near bottom. Axons of ganglion cells form optic nerve. (×100)

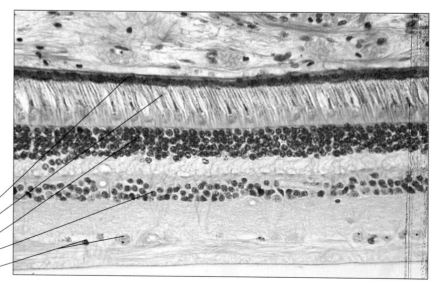

Pigmented epithelium
Rods and cones
Receptor nuclei
Bipolar cell nuclei
Ganglion cells

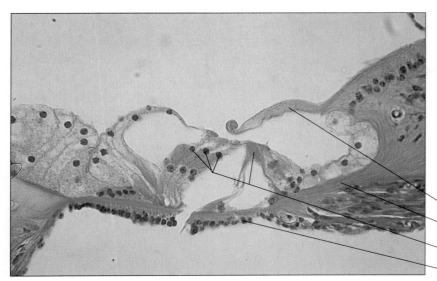

Figure 1-53a
Organ of Corti   Thick finger of tectorial membrane extends from right to stimulate complex of four hair cells (three on left, one on right) of central structure that rests on important basilar membrane. Nerve fibers from hair cells exit right to spiral ganglion for processing and transmission of messages to brain. (×100)

Tectorial membrane

Nerve fibers

Hair cells

Basilar membrane

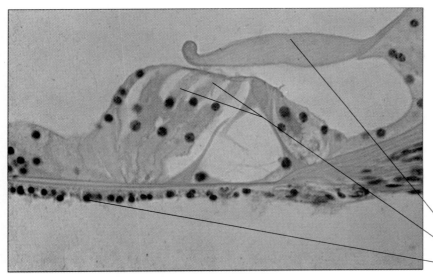

Figure 1-53b
The Organ of Corti   High magnification. (×500)

Tectorial membrane

Hair cells of Organ of Corti

Basilar membrane

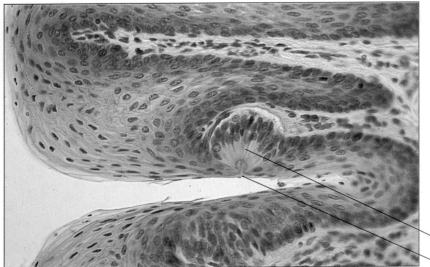

Figure 1-54
Taste Bud   Dissolved chemicals enter fungiform papilla through small pore to directly stimulate sensory cells and initiate taste perception. (×100)

Taste bud

Taste pore

Figure 1-55
Thyroid Gland Follicles   Cuboidal
epithelium surrounds endocrine follicles of
the thyroid gland, the only gland that stores
substantial amounts of its own hormone.
(×100)

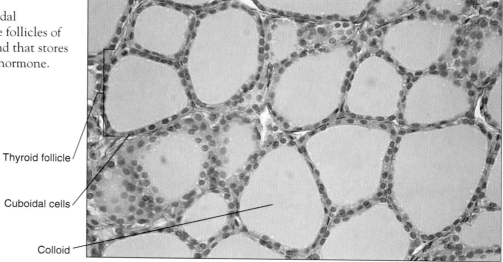

Thyroid follicle

Cuboidal cells

Colloid

Figure 1-56
Parathyroid Gland   LM of section through
the parathyroid gland. (×40)

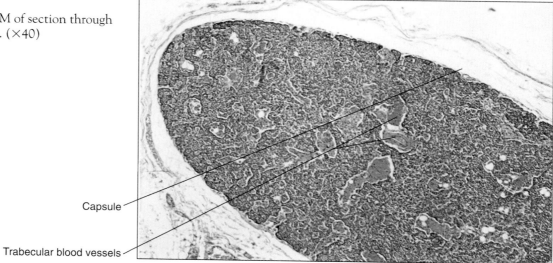

Capsule

Trabecular blood vessels

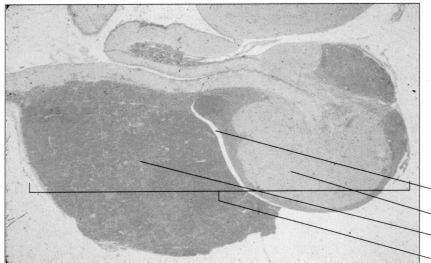

Figure 1-57a
Pituitary Gland   The pituitary gland consists of two components: the posterior component, or neurohypophysis (light stain), consists of mainly nervous tissue, whereas the anterior component, or adenohypophysis (dark stain) consists of a glandular epithelium. (×10)

Cleft

Neurohypophysis

Adenohypophysis

Pituitary gland

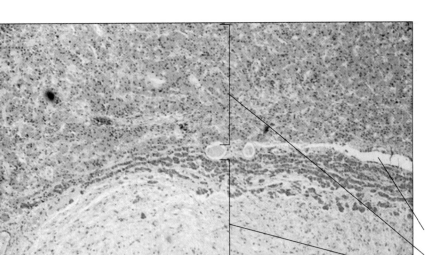

Figure 1-57b
Pituitary Gland   The cleft between the neurohypophysis and adenohypophysis is visible in this view of the pituitary gland. (×100)

Cleft

Adenohypophysis

Neurohypophysis

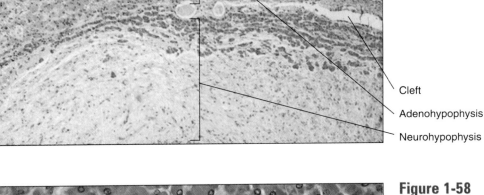
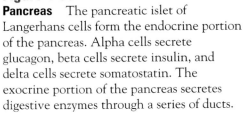
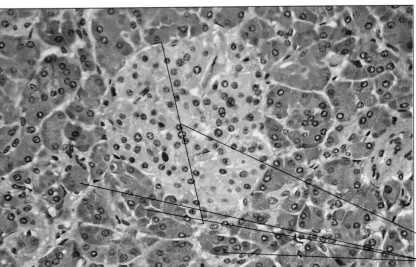

Figure 1-58
Pancreas   The pancreatic islet of Langerhans cells form the endocrine portion of the pancreas. Alpha cells secrete glucagon, beta cells secrete insulin, and delta cells secrete somatostatin. The exocrine portion of the pancreas secretes digestive enzymes through a series of ducts.

Islet of Langerhans

Exocrine cells of pancreas

Figure 1-59

Adrenal Cortex   Outer zone of rounded groups of cells (zona glomerulosa) secretes mineralcorticosteroids (aldosterone). Middle zone of cells appearing in rows (zona fasciculata) secretes glucocorticosteroids. Innermost zone of cells arranged in a meshwork (zona reticularis) secretes mainly androgens. (×50)

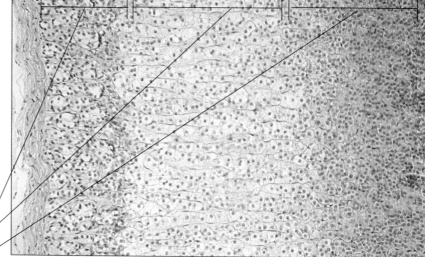

Zona glomerulosa

Zona fasciculata

Zona reticularis

Figure 1-60

Neutrophil   Most numerous (65%) of the leukocytes, it is characterized by a multilobed nucleus and granular cytoplasm. Engages in phagocytosis. (Neutral dyes stain; ×640)

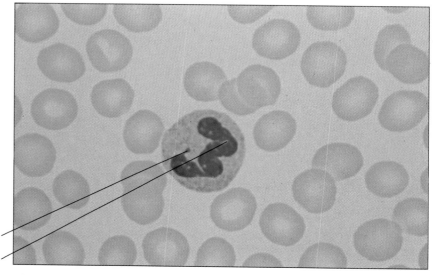

Barr body

Nucleus

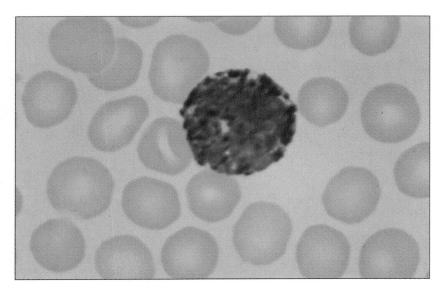

Figure 1-61
Basophil   Normally the rarest (1%) of the leukocytes, its kidney-shaped nucleus may be almost obscured by cytoplasmic granules. These cells contain numerous chemicals involved in inflammation. (Basic dyes stain; ×640)

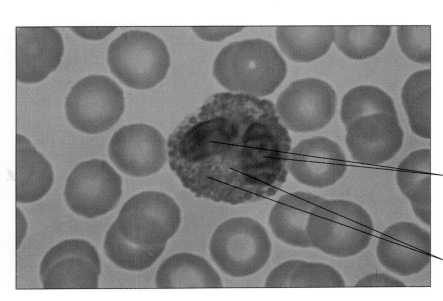

Figure 1-62
Eosinophil   Relatively rare (6%) leukocyte. Usually identifiable because of red-to-orange-staining cytoplasmic granules. Function not definitely known but elevated especially in allergies. (Selective eosin stain; ×640)

— Nucleus (two lobes)

— Granules

Figure 1-63

Lymphocyte   Common (25%). Characterized by single-lobed, "dented" nucleus surrounded by clear cytoplasm. May be large or small. Heavily involved in the immune response including synthesis of antibodies. (×640)

Nucleus

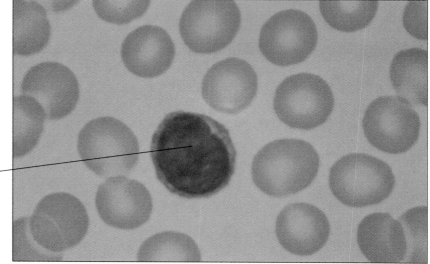

Figure 1-64

Monocyte   Relatively rare (3%). Lobed, often kidney-shaped nucleus is surrounded by clear cytoplasm. Largest of the leukocytes, this cell is a scavenger and engages in phagocytosis. (×640)

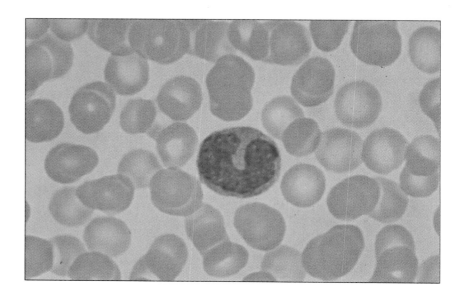

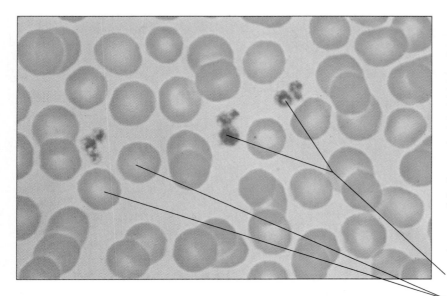

Figure 1-65
Erythrocytes (Red Blood Cells) and Platelets
Circulating erythrocytes are far more common than any of the leukocytes. Normally they have no nucleus but contain the red pigment hemoglobin, which permits them to transport oxygen and carbon dioxide throughout the body. Typically they assume the shape of a biconcave disk. Their diameter of about 7 microns is useful for comparing sizes of other histological structures. Platelets are cellular remnants of a much larger precursor. These remnants contain numerous chemicals, including those important for clotting and inflammation. Platelets initiate blood clotting by forming a plug at wound sites. (×500)

Platelets

Erythrocytes

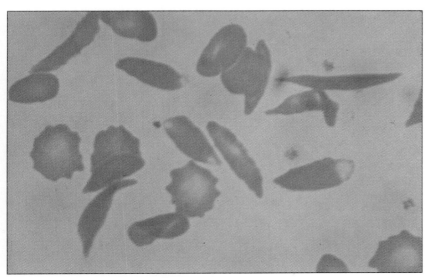

Figure 1-66
Sickle Cell Anemia   Genetic alteration of hemoglobin results in altered membrane structure and abnormal wavy or elongated, curved shape that often resembles a sickle (*upper left*). Oxygen-carrying capacity is much reduced. (×500)

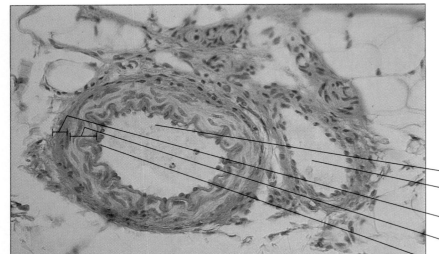

Figure 1-67
Artery (A) and Vein (V)   Blood vessels possess a **tunica intima** that lines the lumen, outside of which is a muscular **tunica media,** and a connective tissue covering, the **tunica adventitia.** The tunica media of arteries is typically much thicker than that of veins. (×100)

A

V   *external*

Tunica adventitia

Tunica media

Tunica intima   *internal*

Figure 1-68a

Arterial Cross Section   Single layer of darkly stained cells, the tunica intima lines the lumen. Thick tunica media is composed of canoe-shaped smooth muscle cells. Outer adventitial layer of connective tissue provides elastic support and strength. (×50)

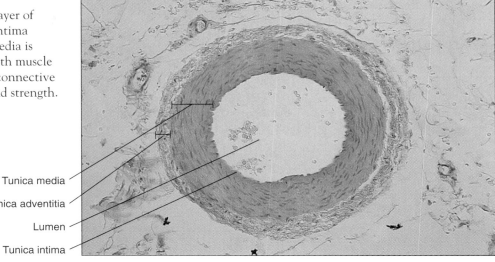

Tunica media

Tunica adventitia

Lumen

Tunica intima

Figure 1-68b

Atherosclerosis   Cross section of a healthy artery.

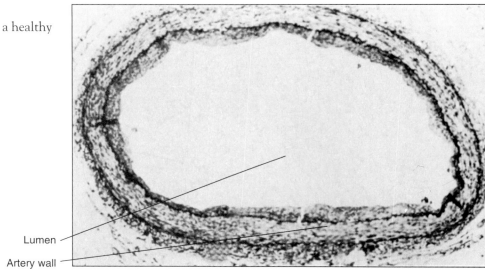

Lumen

Artery wall

Figure 1-68c

Atherosclerosis   Cross section of an artery with advanced atherosclerosis.

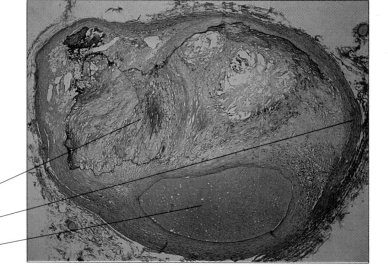

Fatty deposit

Artery wall

Lumen filled with blood

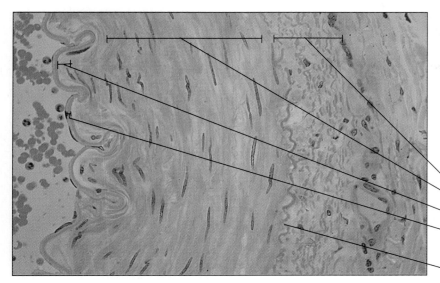

Figure 1-69

Detail of Arterial Wall   Inner endothelial cells of tunica intima (*left*) lie on a basement membrane. A thin layer of smooth muscle cells and elastic tissue (lamina propria) throws this tunic into folds. The tunica media contains multiple layers of smooth muscle cells regularly arranged. A wavy external elastic membrane separates the tunica media from the adventitia.

Adventitia

Tunica media

Lamina propria

Tunica intima

External elastic membrane

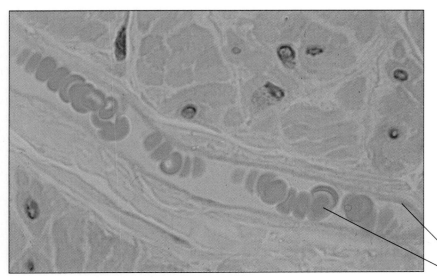

Figure 1-70

Capillary with Red Blood Cells in Single File Capillary wall is made of flattened endothelial cells without complex tunics, a simple structure that facilitates the exchange of gases, nutrients, wastes, and hormones. (×400)

Endothelium

Red blood cell

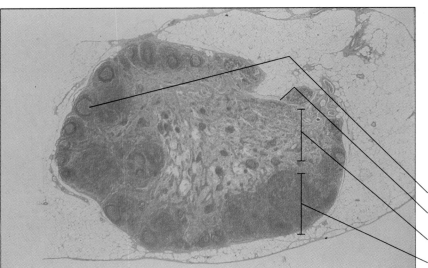

Figure 1-71

Lymph Node   Outer cortex containing several follicles surrounds medulla, with its narrow, dark medullary cords. Notch is hilum, through which blood and lymphatic vessels pass. (×5)

Follicle (germinal center)

Hilum

Medulla

Cortex

Figure 1-72

Valve of Lymphatic Vessel   One-way flow
of lymph, from left to right in this figure, is
ensured by valve action in lymph vessel.
Vessels themselves are thin walled and lack
musculature; pumping action occurs
through compression by neighboring
muscles. (×25)

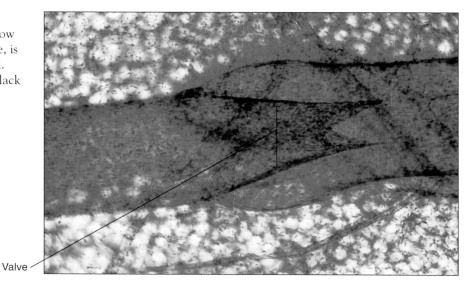

Valve

Figure 1-73a

Thymus   Various lobules contain thick,
darkly staining cortex surrounding a smaller,
lighter-staining medulla. Small, round
cellular patches in medulla are Hassall's
corpuscles. In adults, much of thymus
degenerates and is replaced by adipose
tissue. (×10)

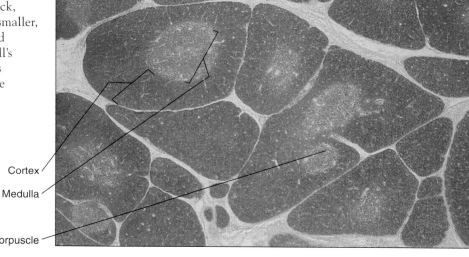

Cortex

Medulla

Hassall's corpuscle

Figure 1-73b

Thymus   Under higher magnification,
the appearance of Hassall's corpuscles
distinguish the thymus from other organs.
Surrounding the corpuscles are reticulate
epithelial cells. (×400)

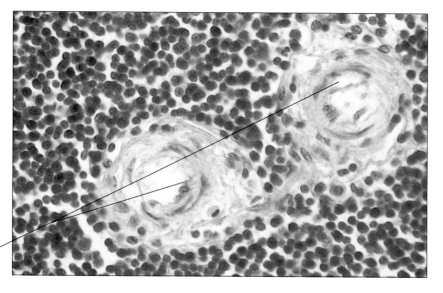

Hassall's (thymic) corpuscles

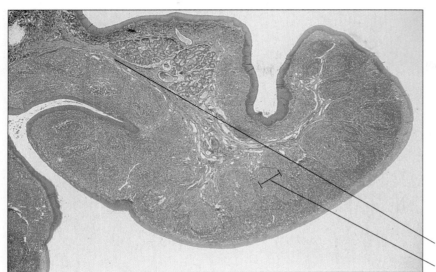

Figure 1-74
Palatine Tonsil   Outer capsule surrounds subcapsular sinus, under which are several large, rounded germinal centers surrounding trabecular arteries and veins. Efferent lymph vessel leads out to upper left. (×5)

Lymph vessel

Germinal center

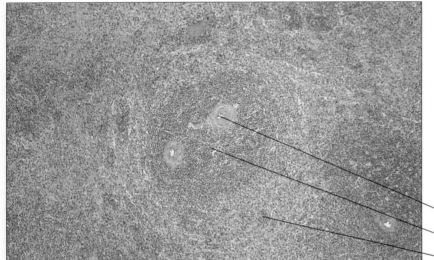

Figure 1-75
Spleen   Central blood vessels are surrounded by area of densely staining white pulp composed of lymphoid cells. Less densely staining red pulp, with fewer cell nuclei, surrounds white pulp. (×25)

Blood vessel

White pulp

Red pulp

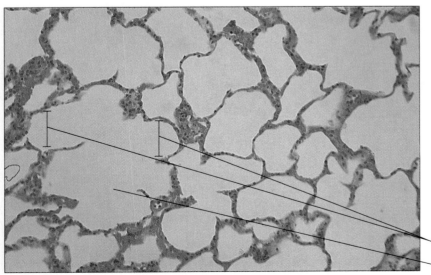

Figure 1-76
Alveoli   Thin-walled respiratory exchange surfaces aid in rapid diffusion of gases. Bronchiole terminates at atrium, which acts as entryway into several individual alveolar sacs, greatly multiplying surface area. (×50)

Alveolar sacs

Atrium

Figure 1-77

Details of Alveolus   Squamous cells compose alveolar sac, which is penetrated by thin-walled blood vessels *(upper left)* containing erythrocytes. (×100)

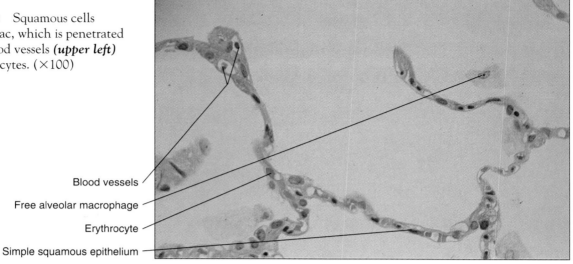

Blood vessels
Free alveolar macrophage
Erythrocyte
Simple squamous epithelium

Figure 1-78

Bronchiole   Epithelial layer that lines the lumen is surrounded by layer of smooth muscle, which regulates bronchiolar diameter. Round structures outside of smooth muscle layer are blood vessels. (×100)

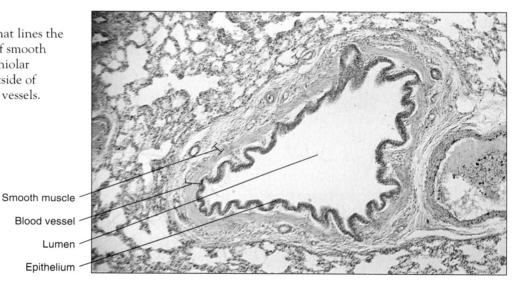

Smooth muscle
Blood vessel
Lumen
Epithelium

Figure 1-79

Esophagus   Surrounding the lumen, esophageal structure contains, in order, the four basic layers of the alimentary canal: **mucosa** (composed of epithelium, the thick lamina propria, and dark muscularis), **submucosa** (light with spaces, blood vessels, and lymph channels), two thick layers of the **muscularis** (circular and longitudinal), and the thin, connective **adventitia** on the surface. Cross section, human. (×3)

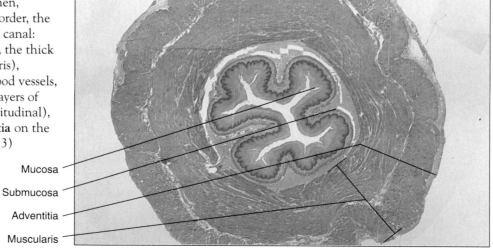

Mucosa
Submucosa
Adventitia
Muscularis

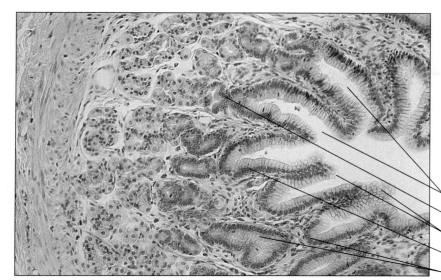

Figure 1-80a

Stomach Mucosa   Visible at entrances to gastric pits are mucus-secreting goblet cells of columnar epithelium. Deeper in pits are acid-secreting parietal cells and enzyme-secreting chief cells. Endocrine-secreting cells near tip of pits are noncolumnar and smaller, with dark, round nuclei. Gastric pits penetrate deep into submucosal layer. Edge of muscularis layer is visible. (×50)

Gastric pits

Endocrine cells

Goblet cells

Parietal cells

Chief cells

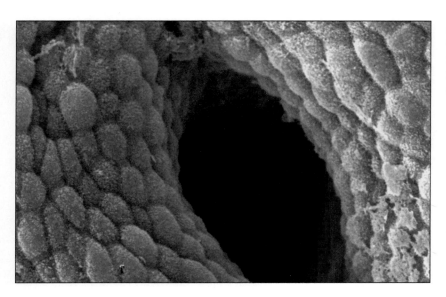

Figure 1-80b

Gastric Pit   The opening of a gastric pit into the stomach, surrounded by the rounded apical surfaces of the columnar epithelial cells of the mucosa.

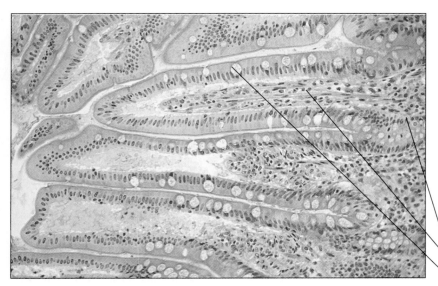

Figure 1-81

Small Intestine, Villi of Ileum (Longitudinal Section)   Numerous pale goblet cells punctuate columnar epithelium that covers each villus. Core of villus contains small blood vessels and blind lymph channel (lacteal). Deep in crypts are endocrine cells, identifiable as dark, round nuclei in a noncolumnar cytoplasm. Human. (×50)

Endocrine cells

Blood vessel and lacteal

Goblet cell

Figure 1-82
Small Intestine, Villi of Ileum (Cross Section)
Goblet cells emptying contents through brush border surface are evident. Core of villus contains blood vessels, lymph channels, and lymphocytes. Human. (×100)

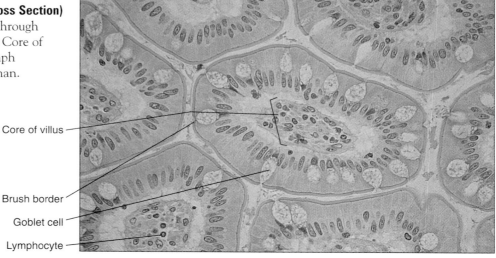

Core of villus

Brush border

Goblet cell

Lymphocyte

Figure 1-83
Large Intestine (Colon) (Cross Section)
Surface is thrown into folds but devoid of villi. Thick submucosa contains blood vessels and lymph channels. (×10)

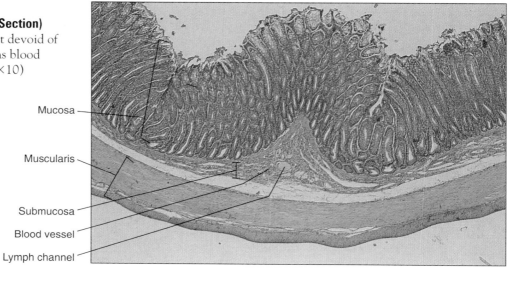

Mucosa

Muscularis

Submucosa

Blood vessel

Lymph channel

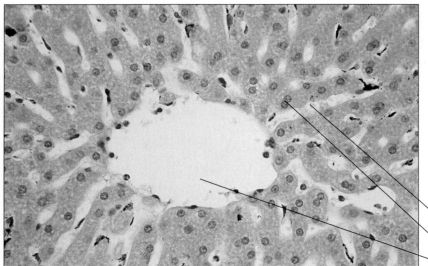

Figure 1-84a
Liver with Central Vein and Sinusoids
Parenchymal hepatocytes lie in radial arrangement around central vein that is lined with single endothelial layer. Cords of hepatocytes are separated by spaces (sinusoids). Sinusoidal surface is covered by microvilli. (×100)

Sinusoid

Hepatocyte

Central vein

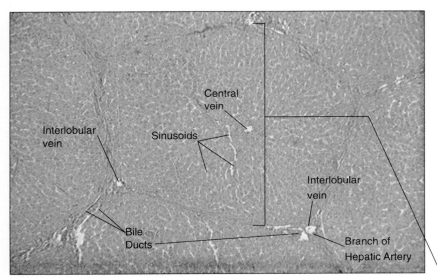

Interlobular vein

Central vein

Sinusoids

Interlobular vein

Bile Ducts

Branch of Hepatic Artery

Lobule

Figure 1-84b
Liver   The liver consists of numerous lobules. A single lobule is in the center of view. At the junction of three adjacent lobes is a bile duct, a branch of the hepatic artery, and a branch of the hepatic portal vein. These three tubes are called a triad. (×40)

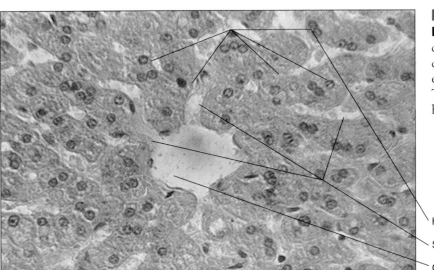

Figure 1-84c
Liver   A single liver lobule consists of a central vein (shown in the center), which collects blood as it flows through narrow endothelial-lined channels, or sinusoids. The cells bordering the sinusoids are called hepatocytes. (×400)

Hepatocytes

Sinusoids

Central vein

Figure 1-85

Gallbladder   Mucosal folds are covered by epithelium with well-developed microvilli. Lamina propria contains blood vessels. (×25)

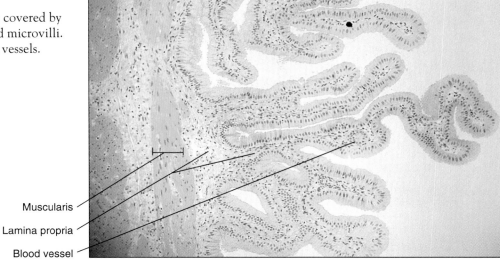

Muscularis

Lamina propria

Blood vessel

Figure 1-86

Vermiform Appendix (Cross Section) Overall structure resembles that of colon. Large, darkly staining structures are lymphoid follicles, the size and number of which decrease with age. Human. (×3)

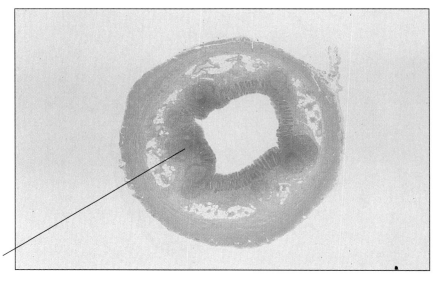

Lymphoid follicle
(germinal center)

Figure 1-87

Sublingual Salivary Gland   Large, pale, mucus-secreting cells, some with caps of serous demilunes, secrete their contents into ducts that may be lined with striated epithelial cells indicative of ion exchange activity. (×100)

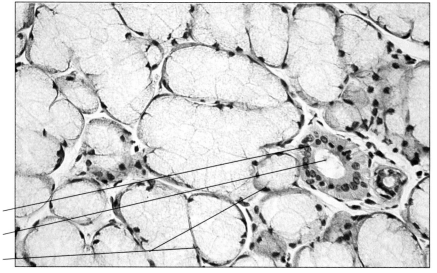

Epithelial cells

Salivary duct

Serous demilunes

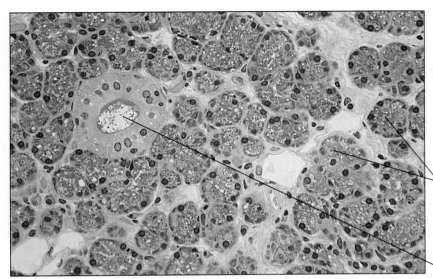

Figure 1-88
Parotid Salivary Gland   Granular serous cells with numerous, large, zymogen granules surround duct. Several tiny ducts run between clusters within the plane of section. Human. (×100)

Zymogen granules

Salivary duct

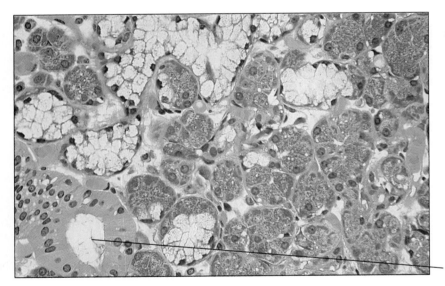

Figure 1-89
Submandibular Salivary Gland with Mucous (Light Staining) and Serous (Dark Staining) Components   Striated duct is visible at lower left. (×100)

Duct

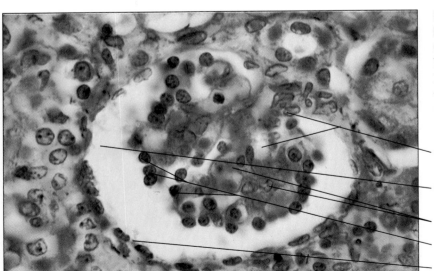

Figure 1-90
Bowman's Capsule and Glomerulus (Renal Corpuscle)   Tuft of capillaries, surrounded by podocytes, protrudes into space of Bowman's capsule. Parietal surface is lined with single layer of simple squamous cells. (×100)

Glomerulus

Space in Bowman's capsule

Podocytes

Glomerulus capsule

Squamous cell

Figure 1-91
Two Glomeruli and Bowman's Capsules
"Lacy" edges of glomerulus on left shows
characteristics of pregnancy-induced
hypertension (PIH), here induced
experimentally in a pregnant rat. (×50)

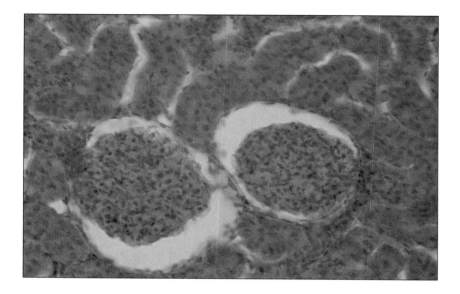

Figure 1-92
**Distal Convoluted Tubules Lined with
Cuboidal Epithelium**   Cross section of rat.
(×400)

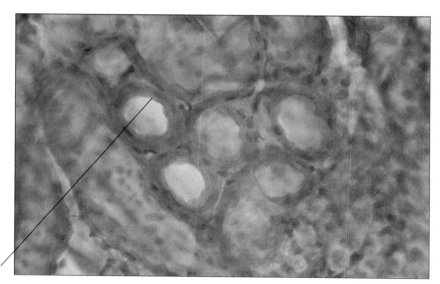

Cuboidal cell

Figure 1-93
Ureter   Star-shaped lumen is lined with
transitional epithelium that varies in
thickness to change shape as lumen
stretches. Delicate lamina propria
separates epithelium from alternating
layers of circular and longitudinal smooth
muscle. (×25)

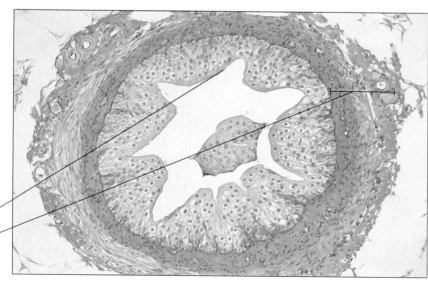

Transitional epithelium

Smooth muscle and adventitial
connective tissue

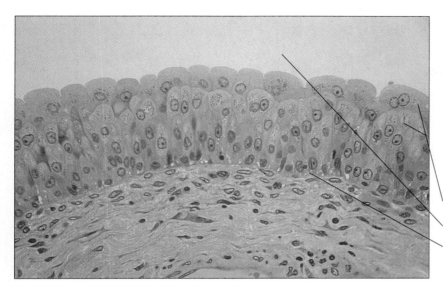

Figure 1-94
Urinary Bladder   Umbrella cells of transitional epithelium stretch and flatten as bladder fills. Basement membrane separates epithelium from underlying connective tissue containing blood vessels. Monkey. (×100)

Umbrella cells

Lumen of bladder

Basement membrane

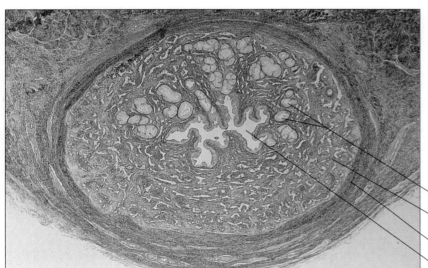

Figure 1-95
Urethra (within Penis)   Lumen is lined with transitional epithelium and is embedded in corpus spongiosum of the penis. Paraurethral glands located above the lumen in the figure secrete mucus into the urethra. A smooth muscle layer (tunica muscularis) surrounds the urethral structures. (×10)

Paraurethral glands

Corpus spongiosum

Tunica muscularis

Lumen

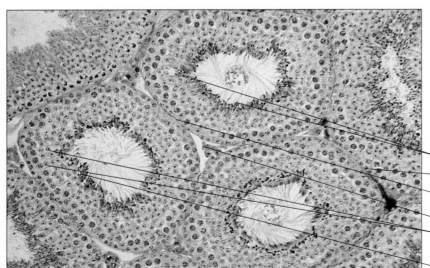

Figure 1-96
Seminiferous Tubules of Testis Lined with Sertoli Cells and Germinativum in Various Stages of Development   Tunica propria surrounds each tubule. Interstitial spaces contain blood vessels and clumps of interstitial (Leydig) cells that secrete testosterone. (×50)

Spermatozoa

Tunica propria

Basement membrane

Interstitial cells

Spermatocytes

Sertoli cells

Figure 1-97
Spermatozoa   Head contains numerous enzymes and nucleus with DNA. Thick midpiece just behind head is packed with mitochondria. (×250)

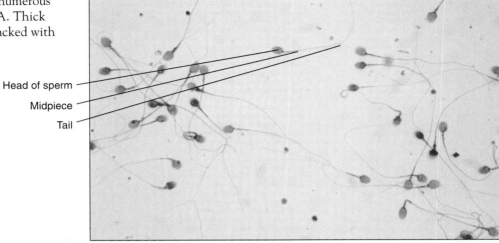

Head of sperm
Midpiece
Tail

Figure 1-98
Epididymis   Tall, pseudostratified columnar epithelium with microvilli surrounds a lumen packed with clumps of spermatozoa. Narrow band of smooth muscle cells encircles each tubule.

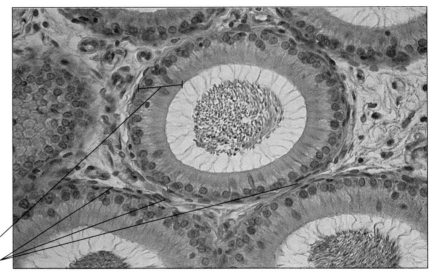

Pseudostratified columnar epithelium
Smooth muscle

Figure 1-99
Ductus Deferens   Ciliated columnar epithelial cells line a spermatozoa-filled lumen. Three layers of smooth muscle cells surround mucosa, a circular layer between two longitudinal ones. (×50)

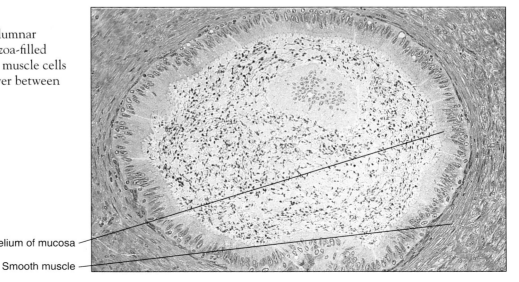

Columnar epithelium of mucosa
Smooth muscle

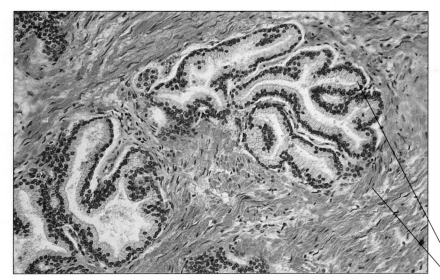

Figure 1-100

Prostate Gland   Mucosal surfaces, lined with tall columnar cells and darkly stained basal nuclei, are arranged in numerous deep folds. Lumina open directly into prostatic urethra. Smooth muscle and fibrocollagenous stroma surround luminal structures. Human. (×50)

Columnar epithelium

Smooth muscle and fibrocollagenous bundles

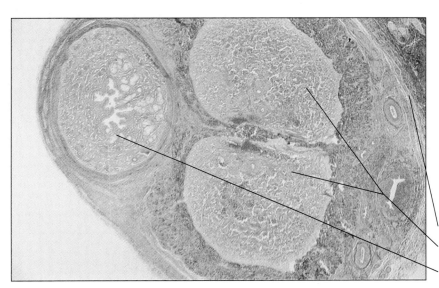

Figure 1-101

Penis   Two corpora cavernosa lie superior to single corpus spongiosum containing penile urethra. Septum between corpora cavernosa is incomplete. Dense fibrous connective tissue, tunica albuginea, surrounds the three vascular cavernosa. The inferior aspect appears on the left, the superior aspect on the right. (×5)

Tunica albuginea

Corpora cavernosa

Corpus spongiosum

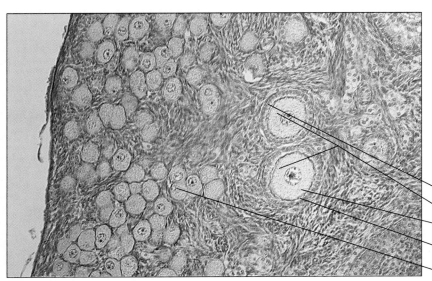

Figure 1-102

Ovary with Numerous Primordial Follicles and Two Primary Follicles   Primordial follicles contain oocytes that are not stimulated to complete the first meiotic division. Two primary follicles each contain an ovum with nucleus and clear surrounding cytoplasm. Thin, clear **zona pellucida** is surrounded by a ring of even cuboidal cells, the **corona radiata.** (×25)

Corona radiata

Primary follicles

Cytoplasm

Membrane of ovum

Primordial follicles

Figure 1-103
Detail of Oocyte in Primordial Follicle
Clear nucleus contains well-defined
nucleolus. Neither zona pellucida nor
corona radiata is evident. (×250)

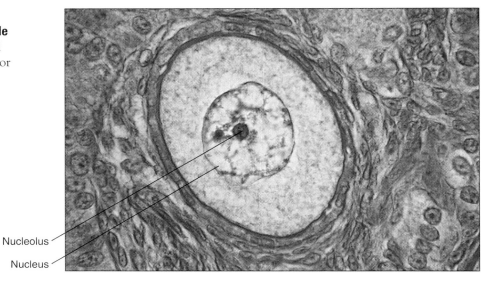

Nucleolus

Nucleus

Figure 1-104
Secondary Ovarian Follicle with Ovum
Bright zona pellucida surrounds outer
membrane of ovum and in turn is
surrounded by dark, cellular corona radiata.
A large **antrum** has formed where the egg
is not anchored to the follicular wall of
granulosa cells. (×100)

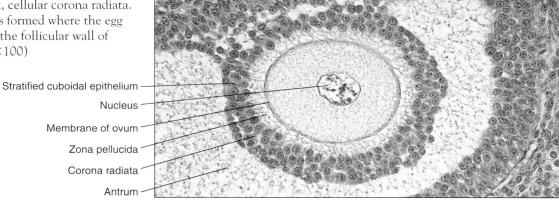

Stratified cuboidal epithelium

Nucleus

Membrane of ovum

Zona pellucida

Corona radiata

Antrum

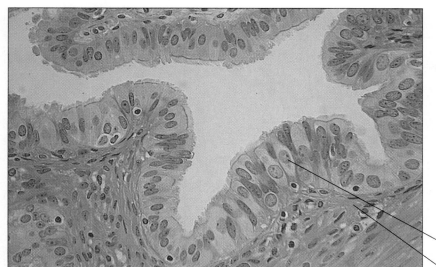

Figure 1-105

Fallopian (Uterine) Tube   Extensive folding of mucosa, lined with ciliated columnar epithelium, is common. Epithelium rests on thin basement membrane and flat connective tissue layer. Rhythmic beating of cilia helps transport ovum toward uterus; cell structure also suggests secretory function. Human. (×100)

Columnar epithelium

Connective tissue

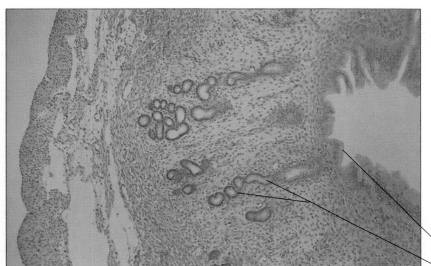

Figure 1-106

Uterus   Endometrial lining (*right*) during proliferative phase of uterine cycle shows thickening of epithelial surfaces and numerous coiled glandular ducts. (×25)

Endothelial lining

Glandular ducts

Human Skeletal Anatomy

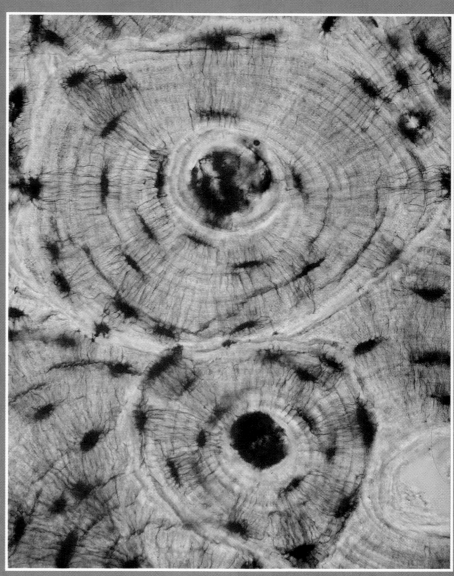

Detail of Compact Bone

Figure 2-1
Skull: Face

BONE

1. Frontal
2. Inferior concha
3. Lacrimal
4. Nasal
5. Parietal
6. Sphenoid
7. Temporal
8. Vomer
9. Zygomatic (malar)
19. Maxilla

FORAMINA

10. Orbital fissure
11. Optic
12. Supraorbital
13. Infraorbital
14. Lacrimal
15. Mental

PROCESSES

16. Mandibular alveolus
17. Maxillary alveolus
18. Perpendicular plate of ethmoid
20. Temporal process of malar
21. Zygomatic process of maxilla
22. Mandibular ramus
23. Frontal notch
24. Anterior nasal spine

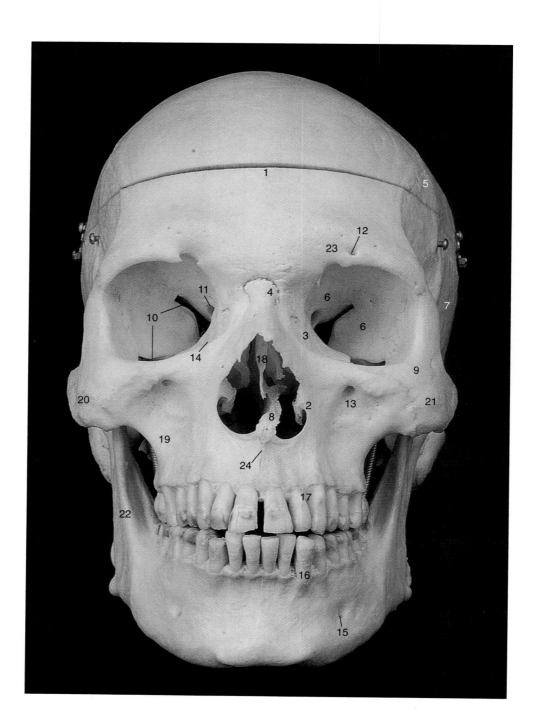

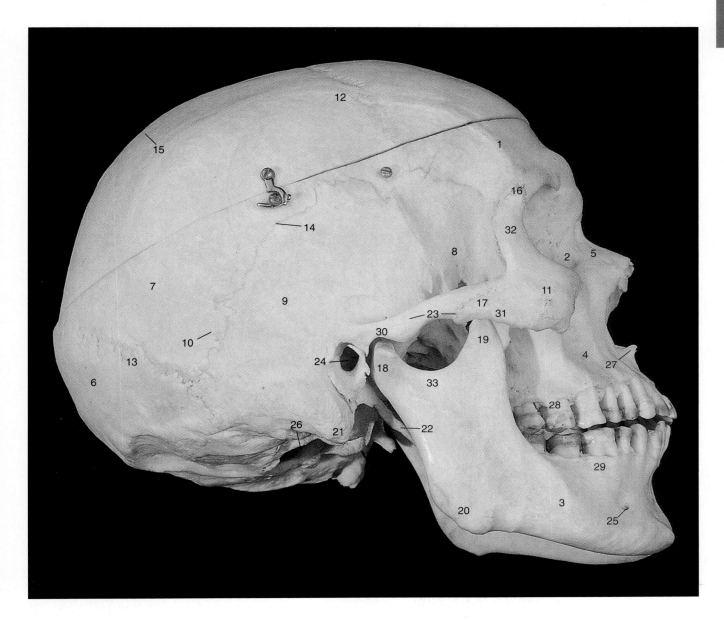

Figure 2-2
Skull: View from Right Side

BONE
1. Frontal
2. Lacrimal
3. Mandible
4. Maxilla
5. Nasal
6. Occipital
7. Parietal
8. Sphenoid (greater wing)
9. Temporal
10. Wormian
11. Zygomatic (malar)

SUTURES
12. Coronal
13. Lambdoidal
14. Squamosal
15. Sagittal
16. Frontozygomatic
17. Temporozygomatic

FORAMINA & PROCESSES
18. Mandibular condyloid process
19. Mandibular coronoid process
21. Mastoid process
22. Styloid process
23. Zygomatic arch

24. External auditory meatus
25. Mental foramen
2. Lacrimal foramen
20. Mandibular angle
26. Foramen magnum
27. Anterior nasal spine
28. Maxillary alveolus
29. Mandibular alveolus
30. Zygomatic process of temporal bone
31. Temporal process of malar
32. Frontal process of malar
33. Mandibular notch

Figure 2-3
Skull: Calvarium, Superior Aspect

BONE
1. Frontal
2. Parietal
3. Occipital

SUTURES
4. Coronal
5. Occipital
6. Sagittal
7. Bregma
8. Lambda

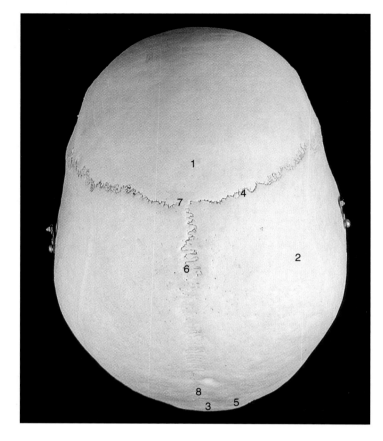

Figure 2-4
Skull: Floor of Cranium, Internal View

BONE
1. Frontal
2. Occipital
3. Parietal
4. Temporal
5. Ethmoid
6. Sphenoid
   Incus
   Malleus } not shown
   Stapes

FORAMINA
7. Foramen magnum
8. Foramen ovale
9. Optic foramen
10. Foramen rotundum
11. Foramen lacerum
12. Carotid canal
13. Jugular fossa and
    foramen

14. Internal auditory
    meatus
15. Grooves for transverse
    and sigmoid sinuses
16. Foramen spinosum

PROCESSES
17. Cribriform plate of
    ethmoid
18. Crista galli
19. Lesser wing of
    sphenoid
20. Greater wing of
    sphenoid
21. Sella turcica
22. Petrous portion of
    temporal bone
23. Orbital plate of
    frontal bone
24. Frontal sinus

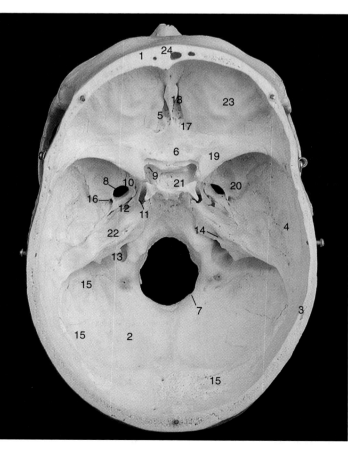

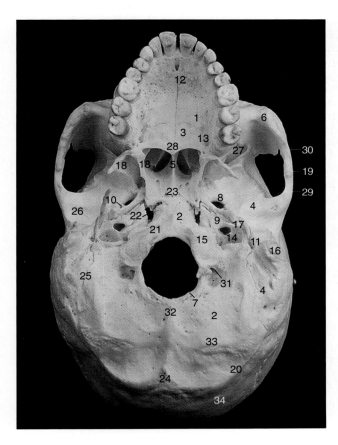

Figure 2-5
Skull: Base, Viewed from Below

BONE

1. Maxilla (palatine process)
2. Occipital
3. Palatine
4. Temporal
5. Vomer
6. Zygomatic (malar)

FORAMINA

7. Foramen magnum
8. Foramen ovale
9. Carotid canal
10. Foramen spinosum
11. Stylomastoid
12. Incisive
13. Palatine
14. Jugular
21. Hypoglossal canal
22. Foramen lacerum
27. Inferior orbital fissure
29. Zygomatic process of temporal bone
30. Temporal process of malar
31. Condylar fossa and canal

PROCESSES

15. Occipital condyle
16. Mastoid
17. Styloid
18. Medial and lateral pterygoid processes
19. Zygomatic arch
20. Superior nuchal line
23. Occipitosphenoid suture
24. External occipital protruberance
25. Occipitotemporal suture
26. Mandibular fossa
28. Posteriornasal spine
32. External occipital crest
33. Inferior nuchal line
34. Highest nuchal line

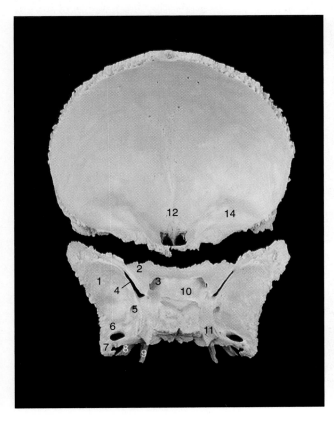

Figure 2-6
Frontal and Sphenoid Bones: Internal Aspect, Frontosphenoidal Suture Separated

PROCESSES

1. Greater wing
2. Lesser wing
8. Lateral pterygoid
9. Medial pterygoid
10. Sella turcica
11. Carotid groove
12. Frontal crest
14. Orbital plate

FORAMINA

3. Optic foramen
4. Superior orbital fissure
5. Foramen rotundum
6. Foramen ovale
7. Foramen spinosum
13. Ethmoidal notch

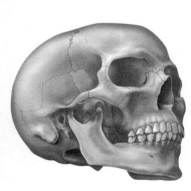
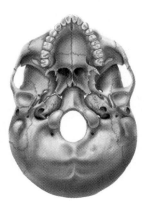

Figure 2-7
Sphenoid: From the Front

PROCESSES

1. Orbital surface
2. Rostrum
7. Medial pterygoid
8. Lateral pterygoid
11. Temporal surface

FORAMINA

3. Optic
4. Superior orbital fissure
5. Foramen rotundum
6. Pterygoid canal
9. Foramen ovale
10. Foramen spinosum

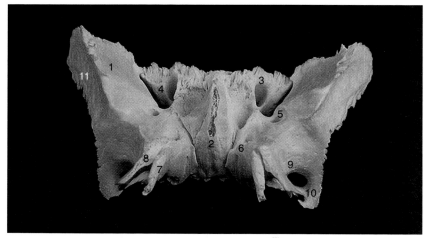

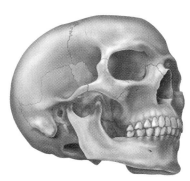

Figure 2-8
Frontal and Zygomatic Bones:
From the Front

PROCESSES

1. Zygomatic process
4. Orbit
7. Frontal process
8. Temporal process
9. Orbital border
10. Maxillary border
11. Temporal border

NOTE: Supraorbital foramen and notch on right side are separate; on left they are superimposed.

FORAMINA

2. Supraorbital foramen
3. Supraorbital notch
5. Ethmoidal notch
6. Infraorbital foramen

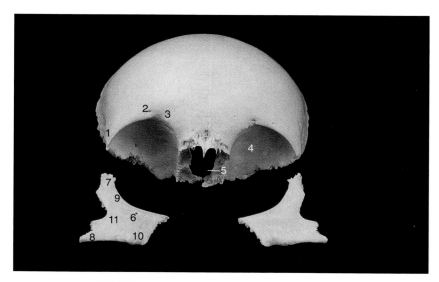

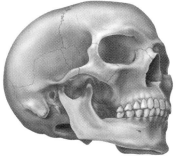

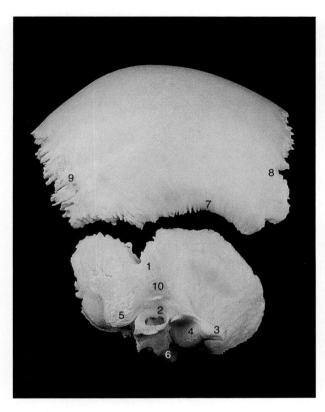

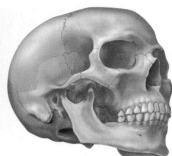

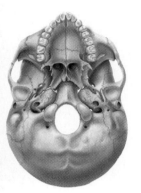

Figure 2-9
Right Temporal and Parietal Bones: Exterior Aspect, Squamosal Suture Separated

PROCESSES

1. Parietal notch
2. External auditory meatus
3. Zygomatic process
4. Mandibular fossa
5. Mastoid process
6. Styloid process
7. Squamosal border
8. Frontal border
9. Occipital border
10. Suprameatal triangle

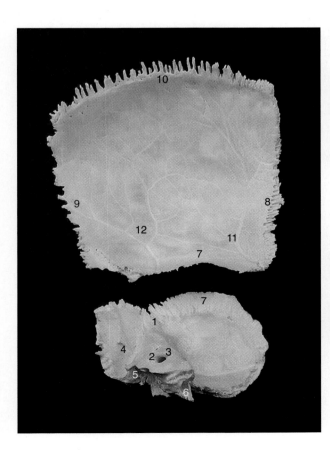

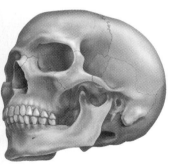

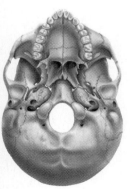

Figure 2-10
Left Temporal and Parietal Bones: Interior Aspect, Squamosal Suture Separated

PROCESSES

1. Parietal notch
2. Internal auditory meatus
3. Petrous portion
4. Groove for sigmoid sinus
5. Mastoid process
6. Styloid process
7. Squamosal border
8. Frontal border
9. Occipital border
10. Sagittal border
11. Furrows for frontal branch of middle meningeal vessels
12. Furrows for parietal branch of middle meningeal vessels

252252222

2222222222

Figure 2-11
Occipital Bone: Interior Aspect

PROCESSES

1. Foramen magnum
2. Condylar fossa and canal
3. Jugular notch
4. Groove for sigmoid sinus
5. Jugular tubercle
6. Mastoid margin
7. Internal occipital crest
8. Internal occipital protuberance
9. Groove for transverse sinus
10. Groove for superior sagittal sinus
11. Lambdoidal margin
12. Cerebellar fossa
13. Cerebral fossa

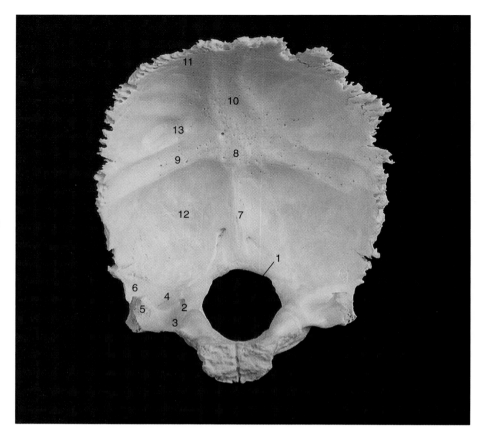

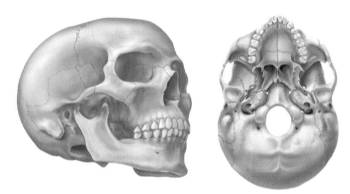

Figure 2-12
Palatine Bones: (A) Right from Anteriomedial View, (B) Left from Anteriosuperior View

PROCESSES

A. Right palatine bone
B. Left palatine bone
1. Orbital process
2. Sphenopalatine notch
3. Sphenoidal process
4. Horizontal plate
5. Ethmoidal crest
6. Pyramidal process
7. Maxillary process
8. Vertical plate

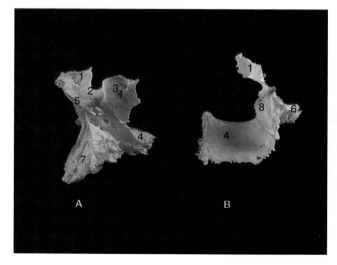

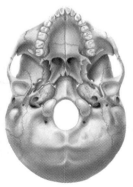

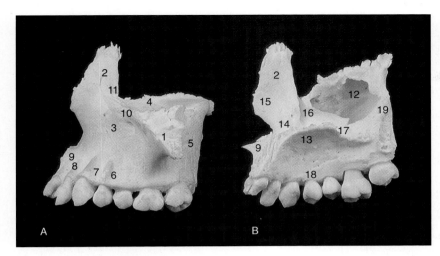

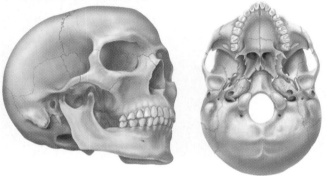

Figure 2-13
**Maxillae: (A) Left from Lateral Aspect,
(B) Right from Medial Aspect**

PROCESSES

A. Left maxilla
B. Right maxilla
1. Zygomatic process
2. Ethmoidal crest
3. Infraorbital foramen
4. Orbital surface
5. Infratemporal surface
6. Canine fossa
7. Canine eminence
8. Incisive fossa
9. Anterior nasal spine
10. Anterior lacrimal crest
11. Nasolacrimal groove
12. Maxillary hiatus and sinus
13. Palatine process
14. Nasal crest
15. Frontal process
16. Middle meatus
17. Inferior meatus
18. Alveolar process
19. Greater palatine canal

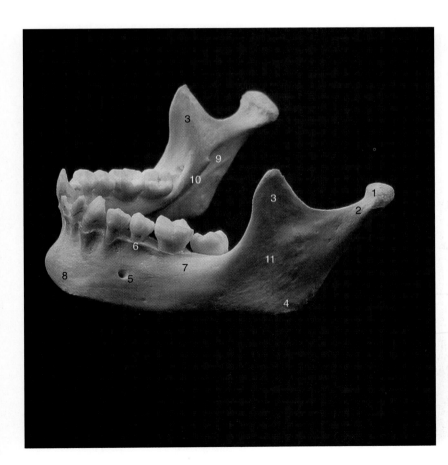

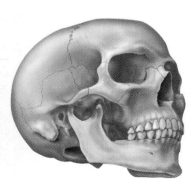

Figure 2-14
Mandible: Left Lateral View

PROCESSES

1. Mandibular condyle
2. Condyloid process
3. Coronoid process
4. Angle
5. Mental foramen
6. Alveolar process
7. Body of mandible
8. Mental protuberance
9. Mandibular foramen
10. Lingula
11. Ramus

Figure 2-15
Ethmoid Bone:
(A) from Above, Right, and Behind
(B) from Below, Right, and Behind

PROCESSES
1. Crista galli
2. Cribriform plate
3. Orbital plate
4. Ethmoidal labyrinth (with air cells)
5. Perpendicular plate
6. Middle nasal concha

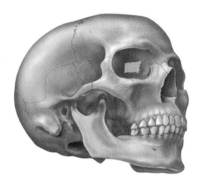

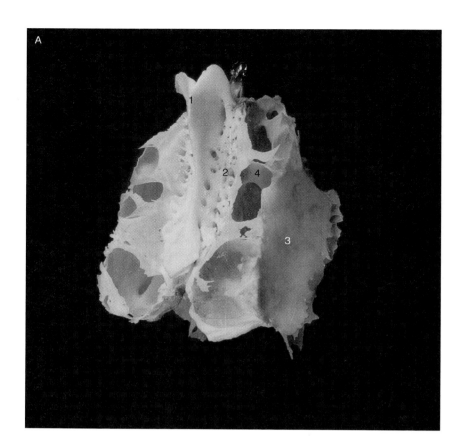

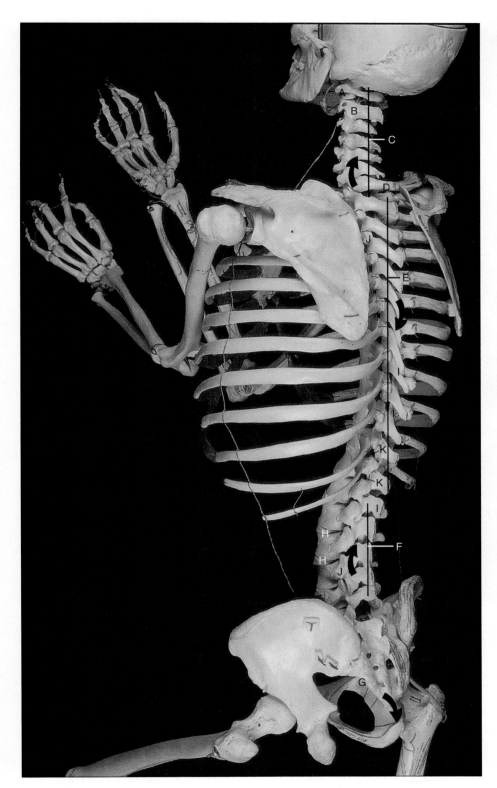

Figure 2-16
Vertebral Column: View from Left and Behind

STRUCTURES

A. Atlas
B. Axis
C. 7 Cervical vertebrae
   (Arrow near C—Cervical curvature)
D. Mentum nuchae
   (spinous process of 7th cervical vertebra)
E. 12 Thoracic vertebrae
   (Arrow near E—Thoracic curvature)
F. 5 Lumbar vertebrae
   (Arrow near F—Lumbar curvature)
G. 5 Fused sacral vertebrae
   (Arrow near G—Sacral curvature)
   (Not shown 4–5 coccyx)
H. Intervertebral disk
I. Spinous processes
J. Transverse processes
K. Intervertebral foramen

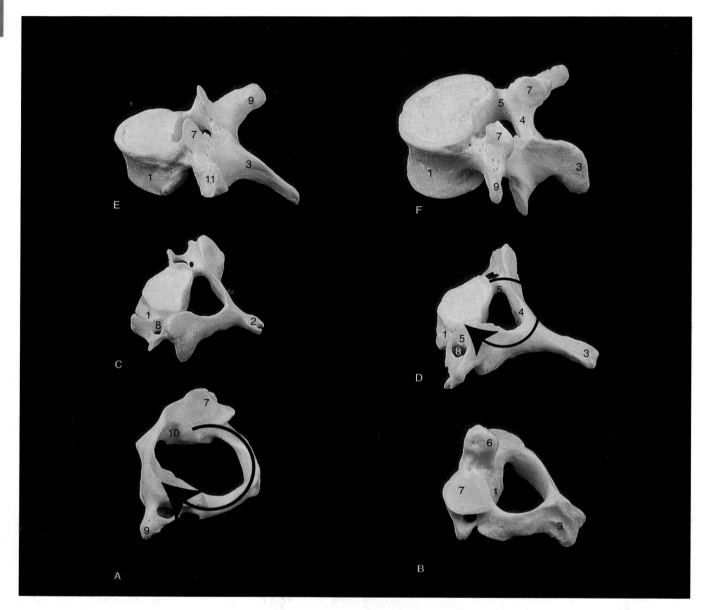

Figure 2-17
Vertebrae

Page 215

PROCESSES

A. Atlas
B. Axis
C. Cervical vertebra
D. 7th cervical vertebra
E. Thoracic vertebra
F. Lumbar vertebra

1. Body
   (Heavy arrows—Vertebral
   arches—comprised of lamina and
   pedicle)
2. Bifid spinous process
3. Monofid spinous process
4. Lamina

5. Pedicle
6. Odontoid process (dens)
7. Superior articular process and facet
8. Transverse foramen
9. Transverse process
10. Vertebral foramen
11. Costal facet

Human Skeletal Anatomy 57

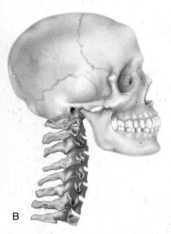

Figure 2-18
(A) Atlas and (B) Axis Articulated

PROCESSES

1. Body of axis
2. Bifid spinous process
3. Odontoid process (dens)
4. Superior articular facet
5. Transverse process
6. Transverse foramen
7. Anterior arch of atlas
   (Heavy arrow—Posterior arch of atlas)
8. Pedicle of axis
9. Lamina of axis

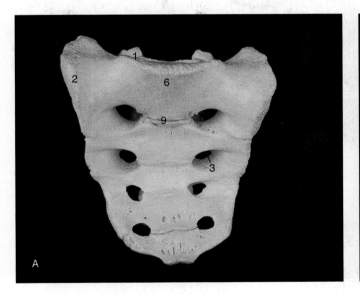

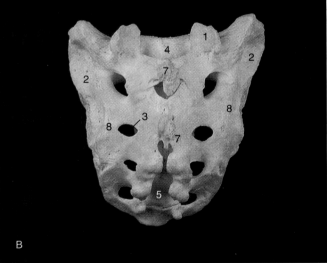

Figure 2-19
Sacrum: (A) Anterior Aspect, (B) Posterior Aspect

PROCESSES

1. Superior articular process and facet
2. Auricular surface
3. Sacral foramen
4. Sacral canal
5. Sacral hiatus
6. Sacral promontory
7. Median crest
8. Lateral crest
9. Site of fusion of 1st and 2nd sacral vertebrae

Figure 2-20
Sternum and Ribs

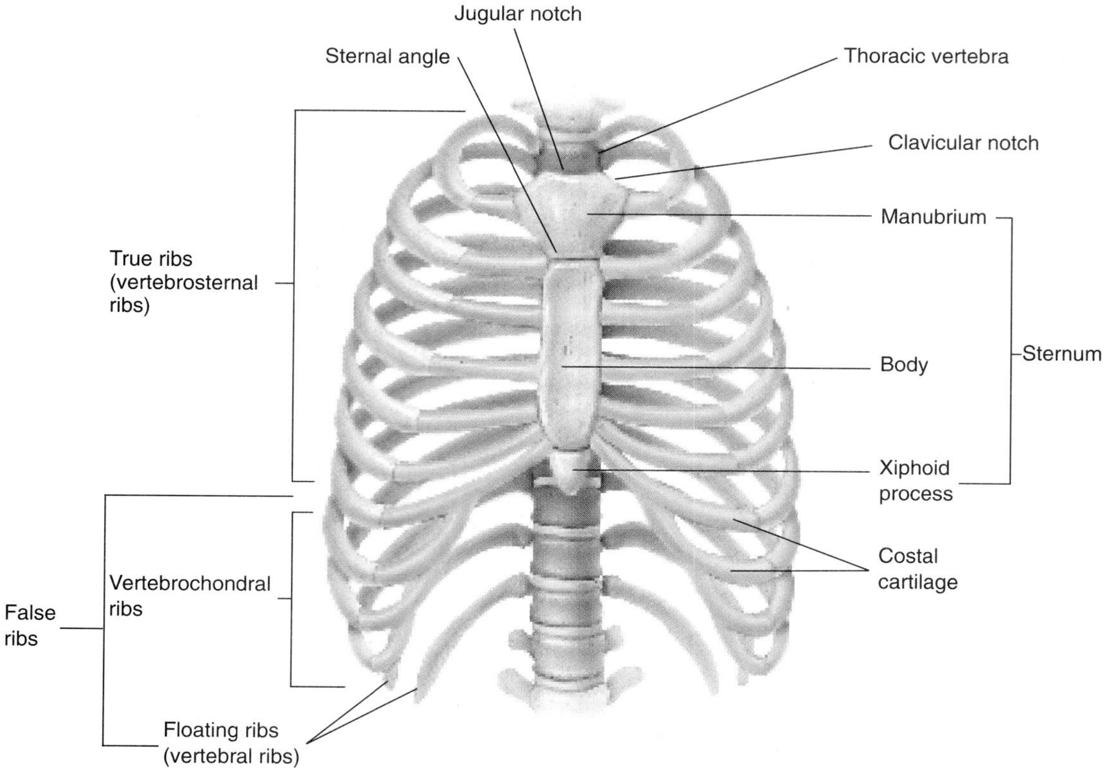

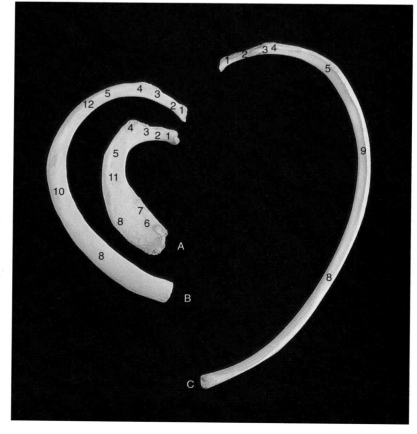

Figure 2-21
Ribs: (A) 1st, (B) 2nd, Right Side, Superior View, (C) 7th, Right Side, Inferior View

PROCESSES

1. Head
2. Neck
3. Articular facet
4. Tubercle
5. Angle
6. Subclavian groove
7. Scalene tubercle
8. Body
9. Costal groove
10. Serratus anterior tuberosity
11. Site for attachment of scalenus medius
12. Site for attachment of levator costa

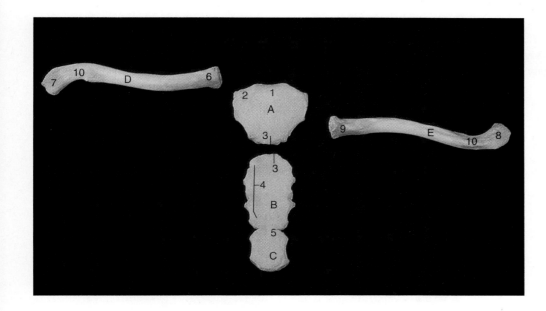

Figure 2-22
Sternum and Clavicles: Sternum from the Front, (D) Right Clavicle from Above, (E) Left Clavicle from Below

PROCESSES

A. Manubrium
B. Gladiolus (body)
C. Xiphoid process
1. Jugular notch
2. Clavicular notch
3. Sternal angle and manubriosternal joint
4. Notches for costal cartilages (2–7)
5. Xiphisternal joint
D. Right clavicle
E. Left clavicle
6. Sternal end
7. Acromial end
8. Conoid tubercle
9. Site for costaclavicular ligament
10. Site for deltoid

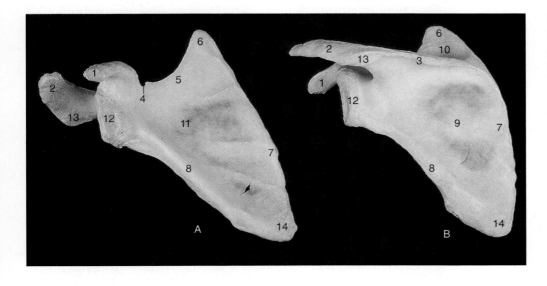

Figure 2-23
Scapulare: (A) Right Scapula, Anterior View (B) Left Scapula, Posterior View

PROCESSES

A. Right scapula
B. Left scapula
1. Coracoid process
2. Acromion
3. Spine
4. Suprascapular notch
5. Superior border
6. Superior angle
7. Medial (vertebral) border
8. Lateral (axillary) border
9. Infraspinous fossa
10. Supraspinous fossa
11. Subscapular fossa
12. Glenoid fossa
13. Acromial angle
14. Inferior angle

Figure 2-24
Humerus with Scapulae:
(A) Right Humerus, Anterior View
(B) Left Humerus, Posterior View

PROCESSES

  A. Right humerus and scapula
  B. Left humerus and scapula
  1. Proximal head (epiphysis)
  2. Anatomical neck
  3. Surgical neck
  4. Shaft (diaphysis)
  5. Lesser tubercle
  6. Greater tubercle
  7. Intertubercular groove
  8. Deltoid tuberosity
  9. Acromion
10. Coracoid process
11. Lateral epicondyle
12. Capitulum
13. Trochlea
14. Medial epicondyle
15. Coronoid fossa
16. Olecranon fossa
17. Distal head and surgical neck

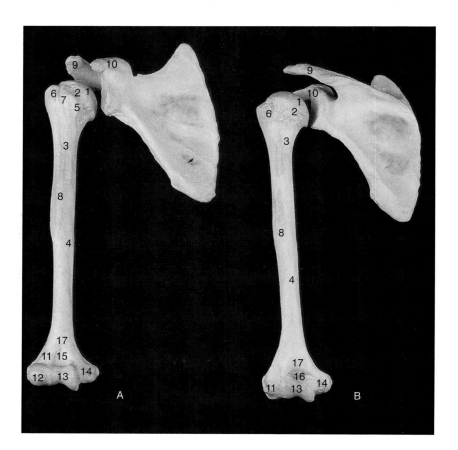

Figure 2-25
Ulna and Radius: Right, Anterior View
Left, Posterior View

PROCESSES

  A. Right radius
  B. Right ulna
  C. Left radius
  D. Left ulna
  1. Proximal head
  2. Neck
  3. Radial tuberosity
  4. Anterior oblique line
  5. Interosseous border
  6. Styloid process of radius
  7. Ulnar notch
  8. Olecranon process
  9. Trochlear notch
10. Coronoid process
11. Ulnar tuberosity
12. Styloid process of ulna
13. Distal head
14. Distal neck
15. Radial notch

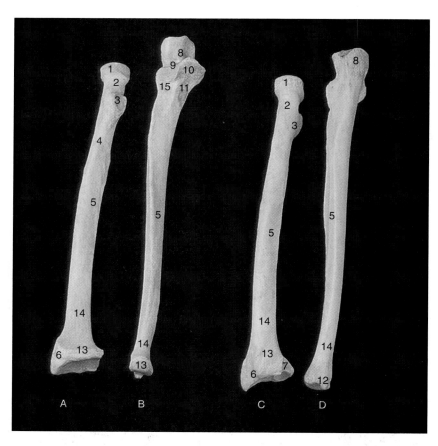

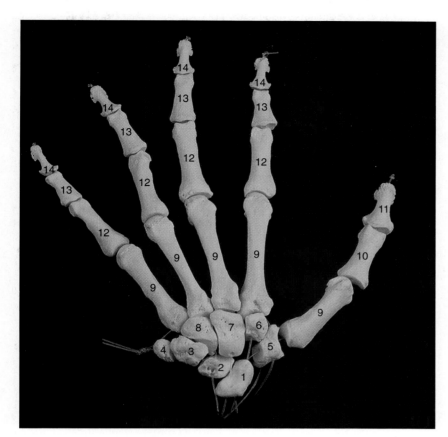

Figure 2-26
Left Hand, Dorsal View

BONES AND PROCESSES

CARPALS
1. Navicular (scaphoid)
2. Lunate
3. Triangular (triquetrum)
4. Pisiform
5. Greater multangular (trapezium)
6. Lesser multangular (trapezoid)
7. Capitate
8. Hamate
9. METACARPALS 1–5

PHALANGES
10. Proximal phalanx of thumb
11. Distal phalanx of thumb
12. Proximal phalanx of digits 2–5
13. Middle phalanx of digits 2–5
14. Distal phalanx of digits 2–5

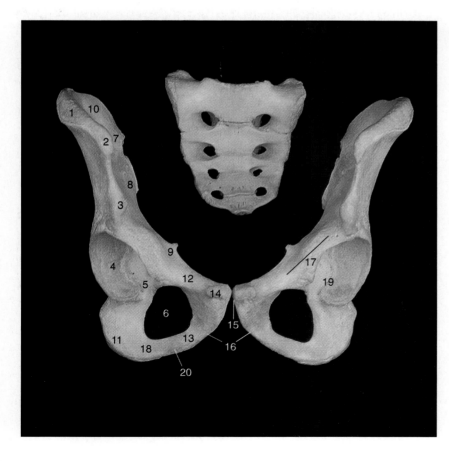

Figure 2-27
Pelvic Bones with Sacrum, Anterior View

PROCESSES
1. Iliac crest
2. Anterior superior iliac spine
3. Anterior inferior iliac spine
4. Acetabulum
5. Acetabular notch
6. Obturator foramen
7. Posterior superior iliac spine
8. Posterior inferior iliac spine
9. Ischial spine
10. Iliac fossa
11. Ischial tuberosity
12. Superior pubic ramus
13. Inferior pubic ramus
14. Pubic tubercle
15. Symphysis pubis
16. Pubic arch
17. Iliopectineal line
18. Ischial ramus
19. Iliopubic eminence
20. Fusion of ischium and pubis

Figure 2-28
Pelvic Bones:
(A) Right Innominate, Lateral View
(B) Left Innominate, Medial View

PROCESSES

1. Iliac crest
2. Anterior superior iliac spine
3. Anterior inferior iliac spine
4. Acetabulum
5. Acetabular notch
6. Obturator foramen
7. Posterior superior iliac spine
8. Posterior inferior iliac spine
9. Ischial spine
10. Iliac fossa
11. Ischial tuberosity
12. Superior pubic ramus
13. Inferior pubic ramus
14. Auricular surface (for sacroiliac joint)
15. Symphysis pubis (cartilage)
16. Greater sciatic notch
17. Iliopectineal line
18. Ischial ramus
19. Iliopubic eminence
20. Fusion of ischium and pubis
21. Lesser sciatic notch

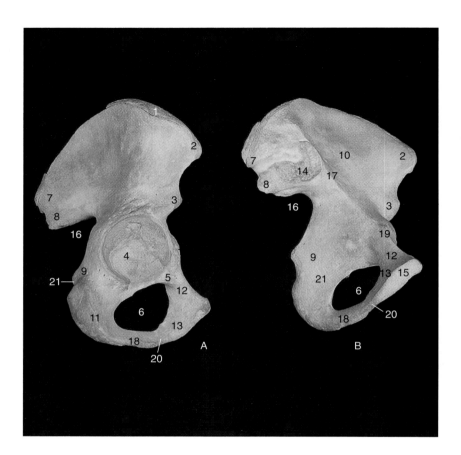

Figure 2-29
Femur: (A) Right Femur, Anterior View
(B) Left Femur, Posterior View

PROCESSES

A. Right femur
B. Left femur
1. Proximal head (epiphysis)
2. Anatomical neck
3. Surgical neck
4. Greater trochanter
5. Lesser trochanter
6. Intertrochanteric crest
7. Spiral line
8. Gluteal tuberosity
9. Linea aspera
10. Lateral condyle
11. Medial condyle
12. Popliteal surface
13. Intercondylar fossa
14. Patellar surface
15. Shaft (diaphysis)
16. Distal head and neck
17. Site for attachment of anterior cruciate
    ligament
18. Site for attachment of posterior cruciate
    ligament

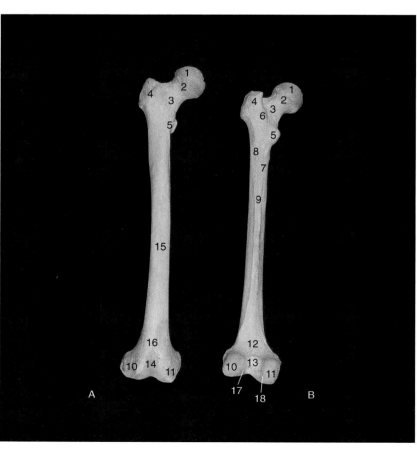

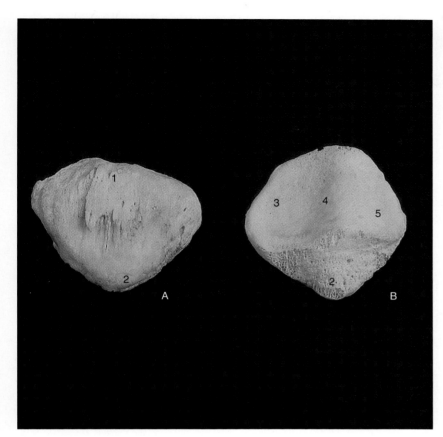

Figure 2-30
**Patellae: (A) Right Patella, Anterior Surface
(B) Left Patella, Posterior (Articular) Surface**

PROCESSES

A. Right patella
B. Left patella
1. Base
2. Apex (site for attachment of patellar ligament)
3. Facet for lateral condyle of femur
4. Vertical ridge
5. Facet for medial condyle of femur

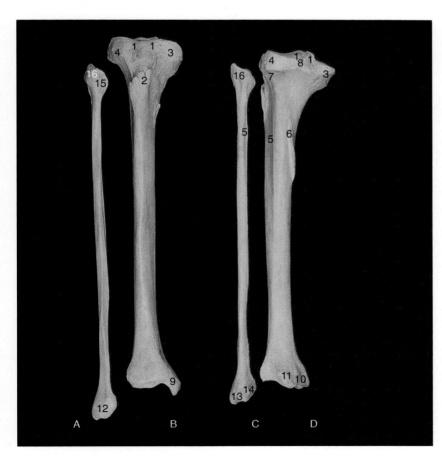

Figure 2-31
**Tibia and Fibula: (A,B) Right Tibia and Fibula, Anterior View
(C,D) Left Tibia and Fibula, Posterior View**

PROCESSES

A. Right fibula
B. Right tibia
C. Left fibula
D. Left tibia
1. Tubercles of intercondylar eminence
2. Tibial tuberosity
3. Medial condyle
4. Lateral condyle
5. Interosseous border
6. Soleal line
7. Articular facet for fibula
8. Site for posterior cruciate ligament
9. Medial malleolus
10. Groove for tibialis posterior
11. Groove for flexor hallicus longus
12. Lateral malleolus
13. Groove for peroneus brevis
14. Malleolar fossa
15. Articular facet for tibia
16. Apex (styloid process)

Figure 2-32
Right Foot: Dorsal (Superior) View

BONES AND PROCESSES

TARSALS
1. Calcaneus
2. Trochlear (articular) surface of talus
3. Lateral tubercle of talus
4. Medial tubercle of talus
5. Groove for flexor hallucis longus
6. Neck of talus
7. Head of talus
8. Navicular
9. Navicular tuberosity
10. Medial cuneiform
11. Intermediate cuneiform
12. Lateral cuneiform
13. Cuboid
14. Metatarsals 1–5

PHALANGES
15. Proximal phalanx of great toe
16. Distal phalanx of great toe
17. Proximal phalanx of digits 2–5
18. Middle phalanx of digits 2–5
19. Distal phalanx of digits 2–5
20. Site for attachment of Achilles tendon

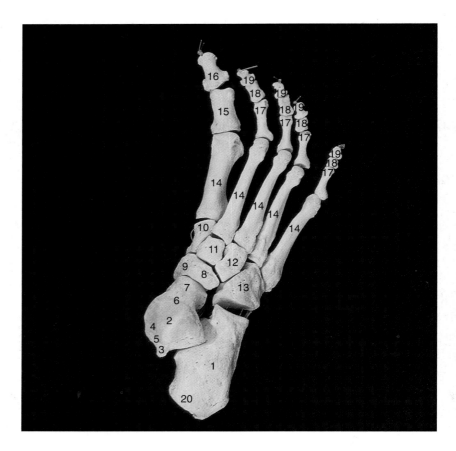

Figure 2-33
Typical Long Bone Structure

PROCESSES
1. Proximal head (epiphysis)
2. Anatomical neck (growth plate, metaphysis, epiphyseal plate)
3. Surgical neck
4. Shaft (diaphysis)
5. Compact bone
6. Cancellous (spongy) bone
7. Distal head
8. Medullary (marrow) cavity

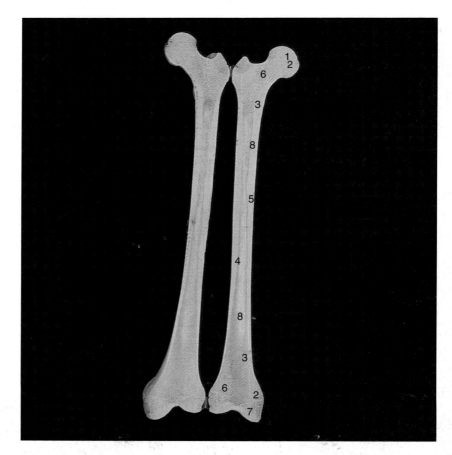

CHAPTER

3

Human Muscular Anatomy

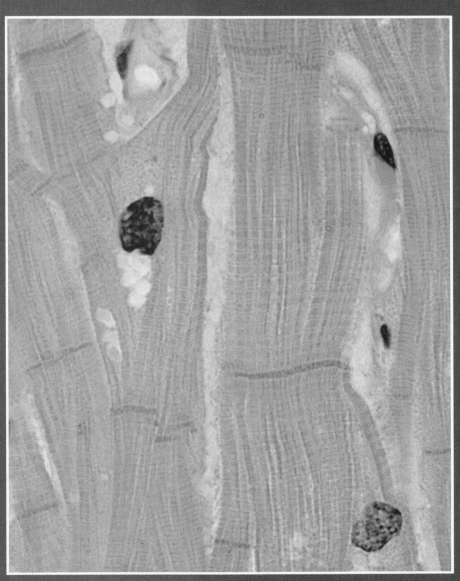

Cardiac Muscle (Longitudinal Section)

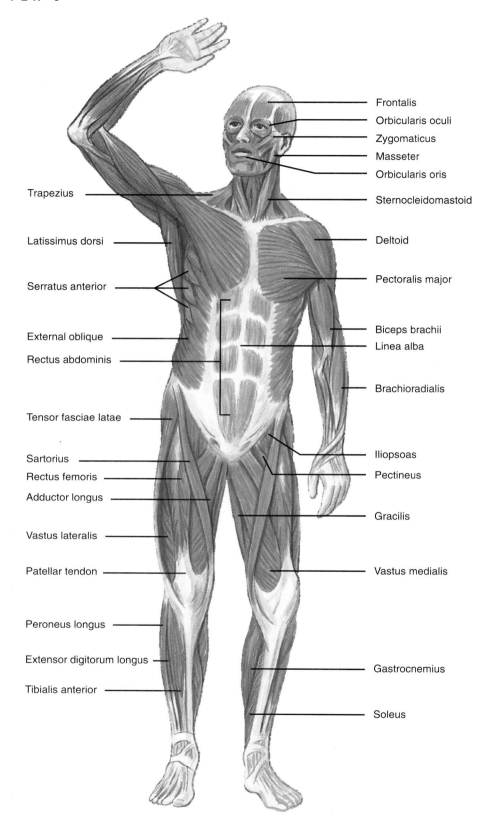

Frontalis

Orbicularis oculi

Zygomaticus

Masseter

Orbicularis oris

Trapezius

Sternocleidomastoid

Latissimus dorsi

Deltoid

Serratus anterior

Pectoralis major

Biceps brachii

External oblique

Linea alba

Rectus abdominis

Brachioradialis

Tensor fasciae latae

Sartorius

Iliopsoas

Rectus femoris

Pectineus

Adductor longus

Vastus lateralis

Gracilis

Patellar tendon

Vastus medialis

Peroneus longus

Extensor digitorum longus

Gastrocnemius

Tibialis anterior

Soleus

Figure 3-1
Superficial Skeletal Muscles
A. Anterior view.

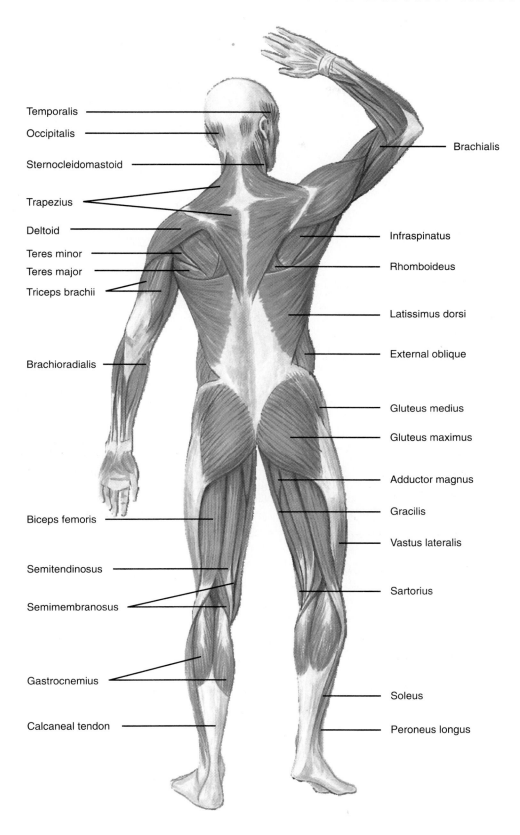

Temporalis

Occipitalis

Sternocleidomastoid

Trapezius

Deltoid

Teres minor

Teres major

Triceps brachii

Brachioradialis

Biceps femoris

Semitendinosus

Semimembranosus

Gastrocnemius

Calcaneal tendon

Brachialis

Infraspinatus

Rhomboideus

Latissimus dorsi

External oblique

Gluteus medius

Gluteus maximus

Adductor magnus

Gracilis

Vastus lateralis

Sartorius

Soleus

Peroneus longus

Figure 3-1—cont'd.
B. Posterior view.

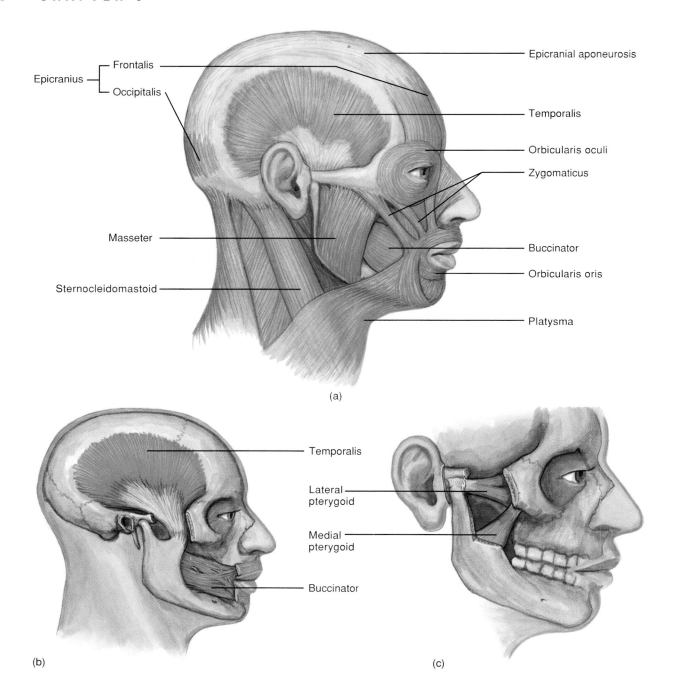

Epicranius
- Frontalis
- Occipitalis

Epicranial aponeurosis

Temporalis

Orbicularis oculi

Zygomaticus

Masseter

Buccinator

Orbicularis oris

Sternocleidomastoid

Platysma

(a)

Temporalis

Lateral pterygoid

Medial pterygoid

Buccinator

(b)

(c)

Figure 3-2
(a) Muscles of facial expression and mastication; isolated views of (b) the temporalis and buccinator muscles and (c) the lateral medial pterygoid muscles.

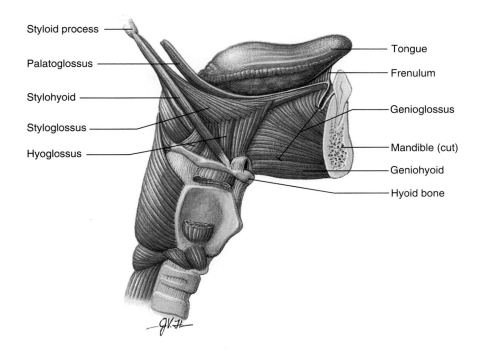

Styloid process

Palatoglossus

Stylohyoid

Styloglossus

Hyoglossus

Tongue

Frenulum

Genioglossus

Mandible (cut)

Geniohyoid

Hyoid bone

Figure 3-3
Muscles of the tongue, right lateral view.

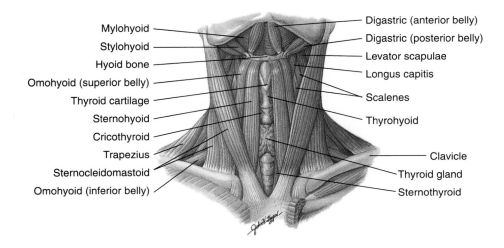

Mylohyoid

Stylohyoid

Hyoid bone

Omohyoid (superior belly)

Thyroid cartilage

Sternohyoid

Cricothyroid

Trapezius

Sternocleidomastoid

Omohyoid (inferior belly)

Digastric (anterior belly)

Digastric (posterior belly)

Levator scapulae

Longus capitis

Scalenes

Thyrohyoid

Clavicle

Thyroid gland

Sternothyroid

Figure 3-4
Muscles of the anterior neck, superficial view.

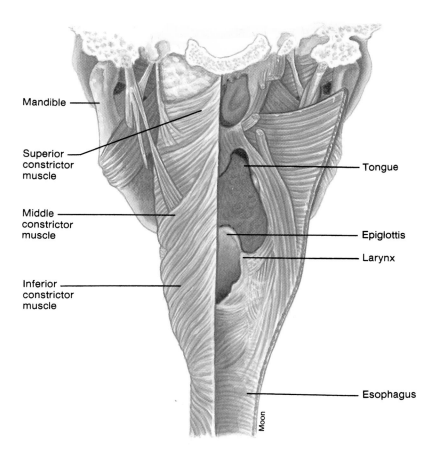

Mandible

Superior
constrictor
muscle

Middle
constrictor
muscle

Inferior
constrictor
muscle

Tongue

Epiglottis

Larynx

Esophagus

Moon

Figure 3-5
Posterior view of the constrictor muscles of the pharynx.
Right side has been cut away to illustrate the interior
structures in the pharynx.

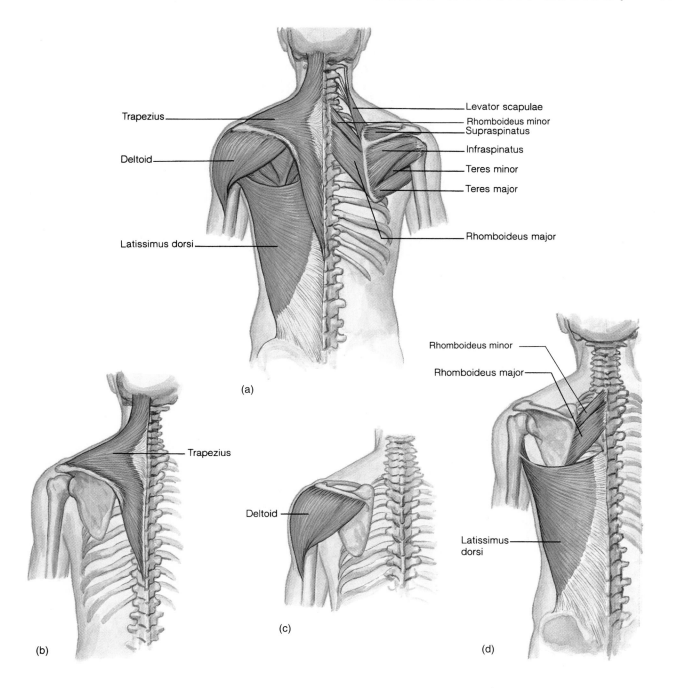

Trapezius

Deltoid

Latissimus dorsi

Levator scapulae
Rhomboideus minor
Supraspinatus
Infraspinatus
Teres minor
Teres major

Rhomboideus major

(a)

Trapezius

(b)

Deltoid

(c)

Rhomboideus minor

Rhomboideus major

Latissimus dorsi

(d)

Figure 3-6

(a) Muscles of the posterior shoulder. The right trapezius is
removed to show underlying muscles. Isolated views of (b)
trapezius, (c) deltoid, and (d) rhomboideus and latissimus
dorsi muscles.

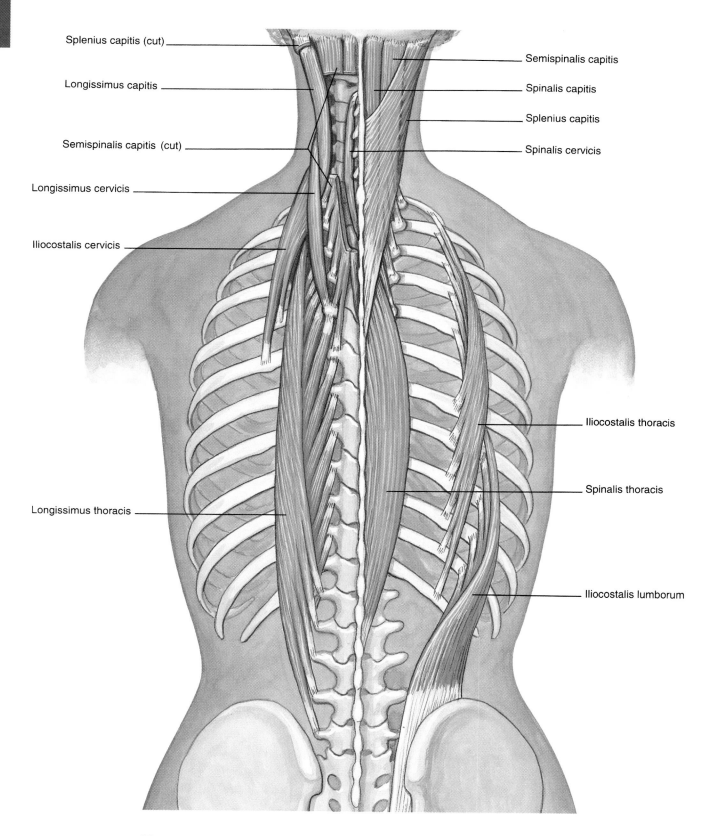

Splenius capitis (cut)

Longissimus capitis

Semispinalis capitis (cut)

Longissimus cervicis

Iliocostalis cervicis

Longissimus thoracis

Semispinalis capitis

Spinalis capitis

Splenius capitis

Spinalis cervicis

Iliocostalis thoracis

Spinalis thoracis

Iliocostalis lumborum

Figure 3-7
Deep muscles of the back and the neck help move the head (posterior view) and
hold the torso erect. The splenius capitis and semispinalis capitis are removed on
the left to show underlying muscles.

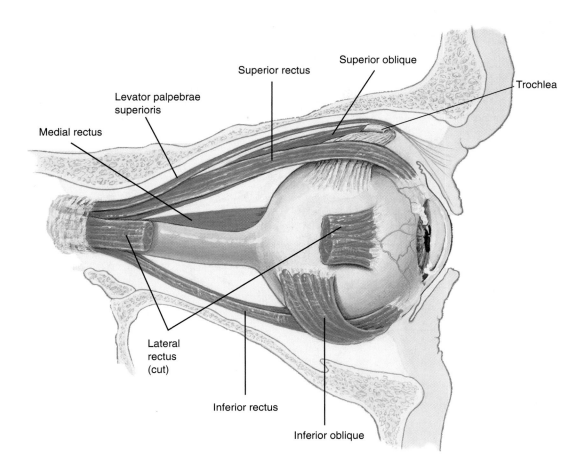

Figure 3-8
Extrinsic muscles of the right eye, lateral view.

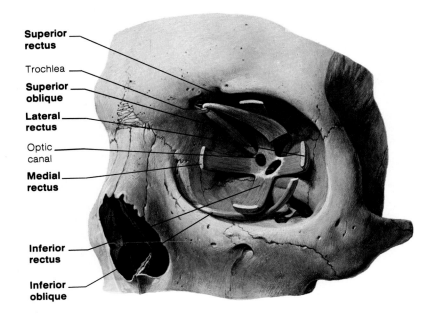

Figure 3-9
Extrinsic muscles of the left eye, anterior view.

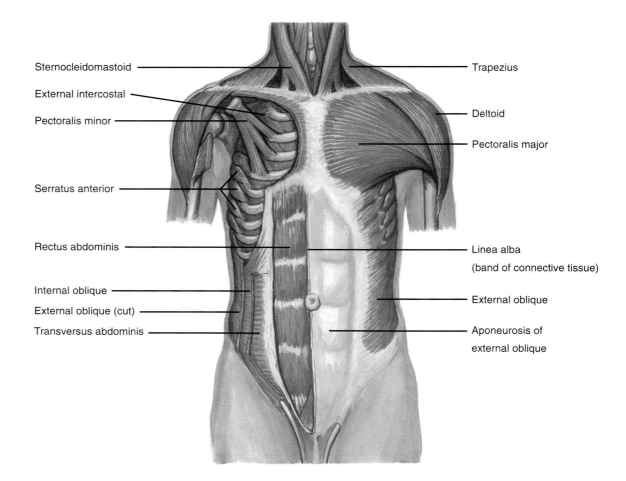

Sternocleidomastoid

External intercostal

Pectoralis minor

Serratus anterior

Rectus abdominis

Internal oblique

External oblique (cut)

Transversus abdominis

Trapezius

Deltoid

Pectoralis major

Linea alba
(band of connective tissue)

External oblique

Aponeurosis of
external oblique

Figure 3-10
Muscles of the anterior chest and abdominal wall. The right
pectoralis major is removed to show the pectoralis minor.

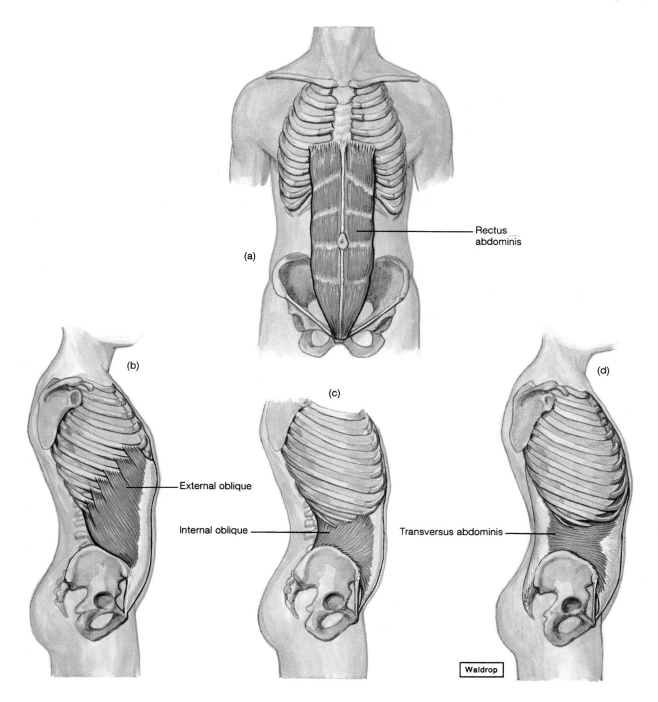

Figure 3-11
(a–d) Isolated muscles of the abdominal wall.

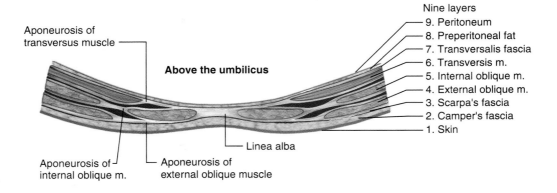

Aponeurosis of
transversus muscle

Above the umbilicus

Nine layers
9. Peritoneum
8. Preperitoneal fat
7. Transversalis fascia
6. Transversis m.
5. Internal oblique m.
4. External oblique m.
3. Scarpa's fascia
2. Camper's fascia
1. Skin

Linea alba

Aponeurosis of
internal oblique m.

Aponeurosis of
external oblique muscle

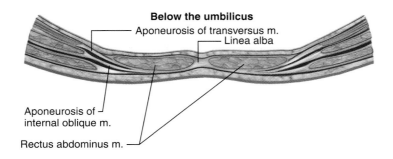

Below the umbilicus
Aponeurosis of transversus m.
Linea alba

Aponeurosis of
internal oblique m.

Rectus abdominus m.

Figure 3-12
Muscles of the anterior abdominal wall, cross-sectional
view above the umbilicus.

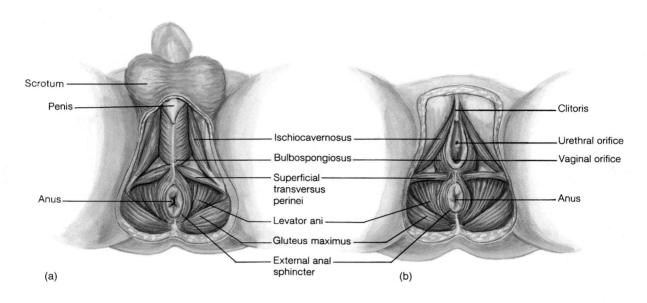

Scrotum

Penis

Ischiocavernosus

Bulbospongiosus

Superficial transversus perinei

Anus

Levator ani

Gluteus maximus

External anal sphincter

(a)

Clitoris

Urethral orifice

Vaginal orifice

Anus

(b)

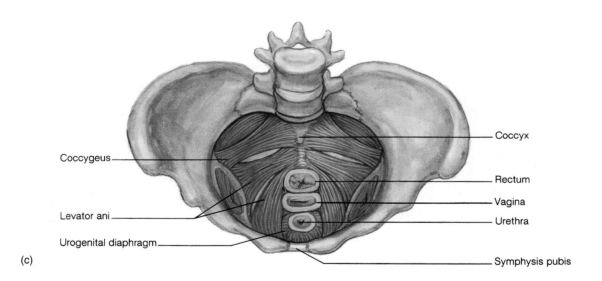

Coccygeus

Levator ani

Urogenital diaphragm

(c)

Coccyx

Rectum

Vagina

Urethra

Symphysis pubis

Figure 3-13

External view of muscles of (a) the male pelvic outlet, and (b) the female pelvic outlet. (c) Internal view of female pelvic and urogenital diaphragms.

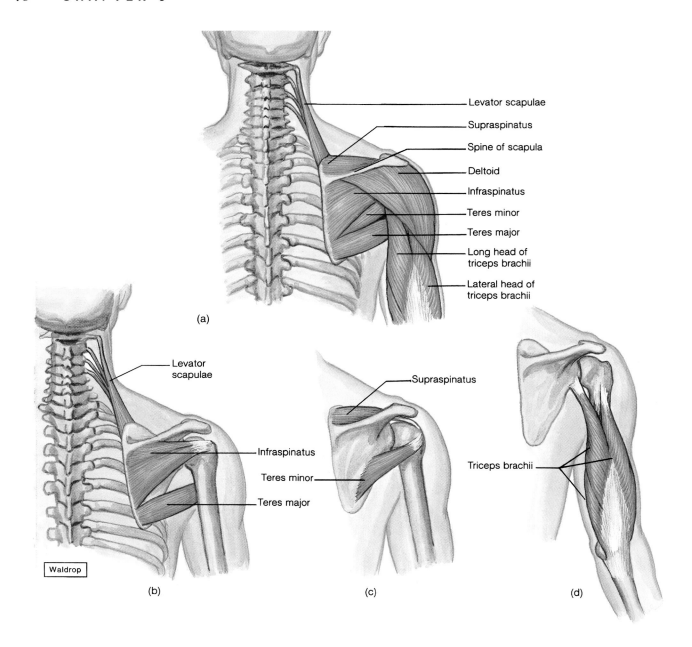

Figure 3-14
(a) Muscles of the posterior surface of the scapula and the arm; (b and c) muscles associated with the scapula. (d) Isolated view of the triceps brachii.

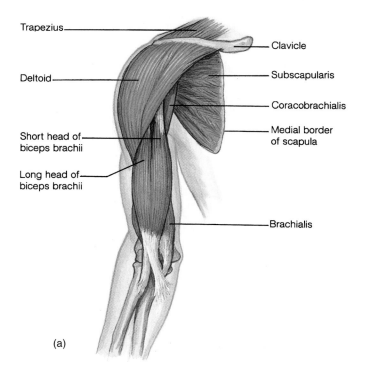

Trapezius

Clavicle

Deltoid

Subscapularis

Coracobrachialis

Medial border
of scapula

Short head of
biceps brachii

Long head of
biceps brachii

Brachialis

(a)

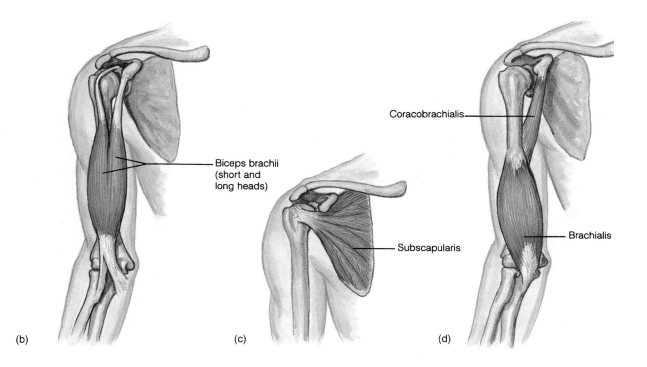

Biceps brachii
(short and
long heads)

Coracobrachialis

Subscapularis

Brachialis

(b)

(c)

(d)

Figure 3-15
(a) Muscles of the anterior shoulder and the arm, with the rib cage removed. (b, c, and d) Isolated views of muscles associated with the arm.

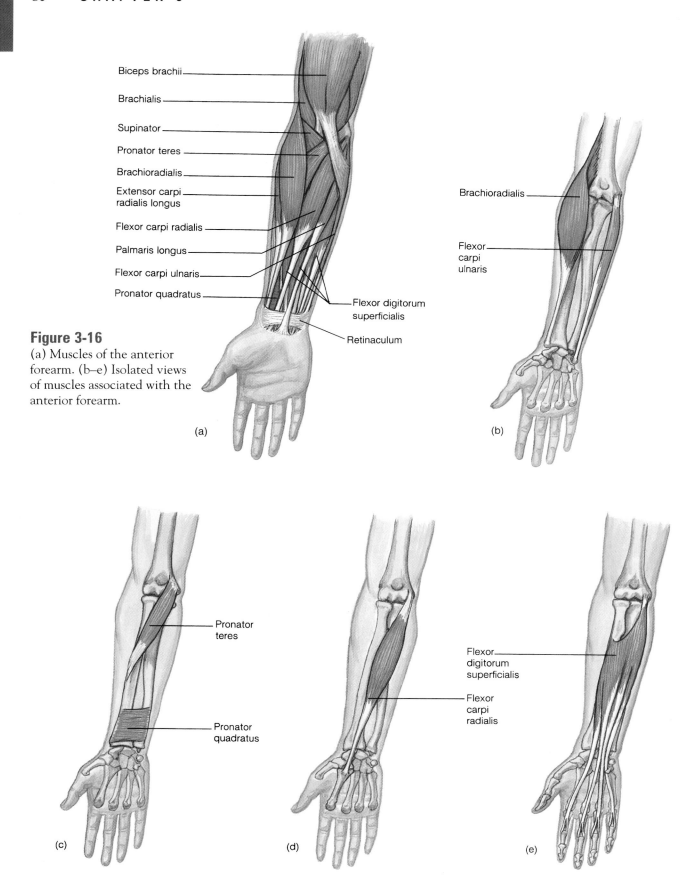

Biceps brachii

Brachialis

Supinator

Pronator teres

Brachioradialis

Extensor carpi radialis longus

Flexor carpi radialis

Palmaris longus

Flexor carpi ulnaris

Pronator quadratus

Flexor digitorum superficialis

Retinaculum

Figure 3-16
(a) Muscles of the anterior forearm. (b–e) Isolated views of muscles associated with the anterior forearm.

(a)

Brachioradialis

Flexor carpi ulnaris

(b)

Pronator teres

Pronator quadratus

(c)

Flexor digitorum superficialis

Flexor carpi radialis

(d)

(e)

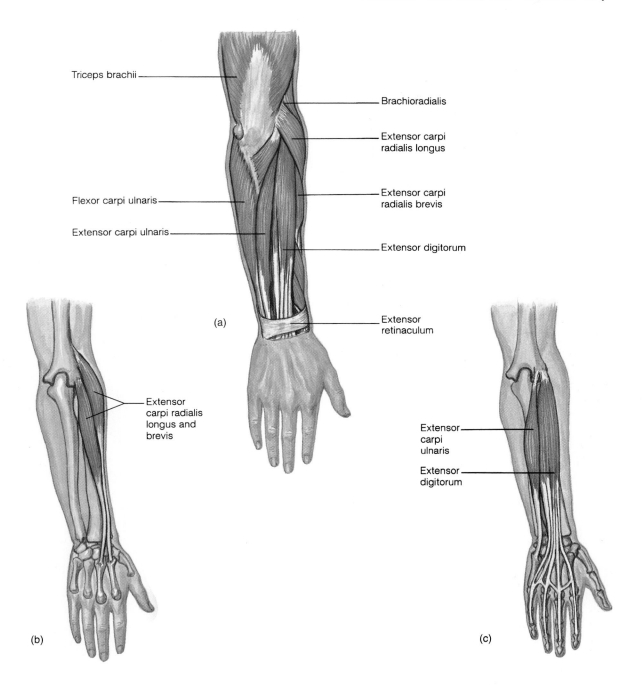

Triceps brachii

Brachioradialis

Extensor carpi radialis longus

Flexor carpi ulnaris

Extensor carpi radialis brevis

Extensor carpi ulnaris

Extensor digitorum

(a)

Extensor retinaculum

Extensor carpi radialis longus and brevis

(b)

Extensor carpi ulnaris

Extensor digitorum

(c)

Figure 3-17
(a) Muscles of the posterior forearm. (b and c) Isolated views of muscles associated with the posterior forearm.

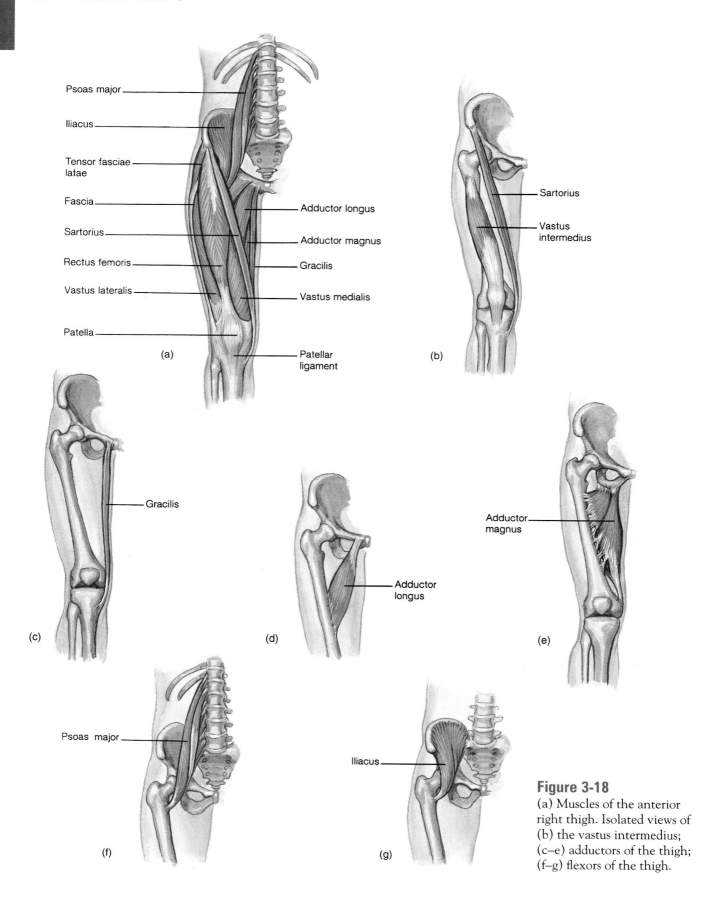

Psoas major

Iliacus

Tensor fasciae latae

Fascia

Sartorius

Rectus femoris

Vastus lateralis

Patella

(a)

Adductor longus

Adductor magnus

Gracilis

Vastus medialis

Patellar ligament

Sartorius

Vastus intermedius

(b)

Gracilis

(c)

Adductor longus

(d)

Adductor magnus

(e)

Psoas major

(f)

Iliacus

(g)

Figure 3-18
(a) Muscles of the anterior right thigh. Isolated views of (b) the vastus intermedius; (c–e) adductors of the thigh; (f–g) flexors of the thigh.

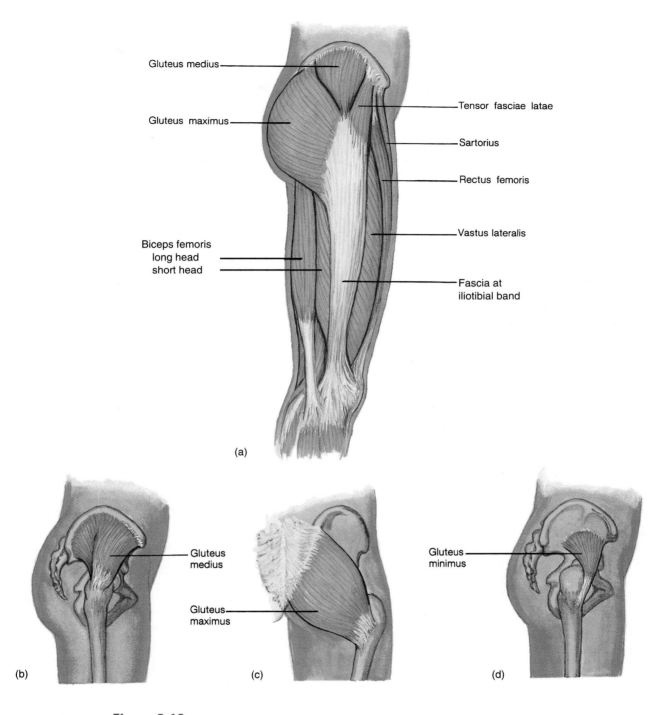

Gluteus medius

Gluteus maximus

Tensor fasciae latae

Sartorius

Rectus femoris

Vastus lateralis

Biceps femoris
long head
short head

Fascia at
iliotibial band

(a)

Gluteus
medius

Gluteus
maximus

Gluteus
minimus

(b)   (c)   (d)

Figure 3-19
(a) Muscles of the lateral right thigh. (b–d) Isolated views of the gluteal muscles.

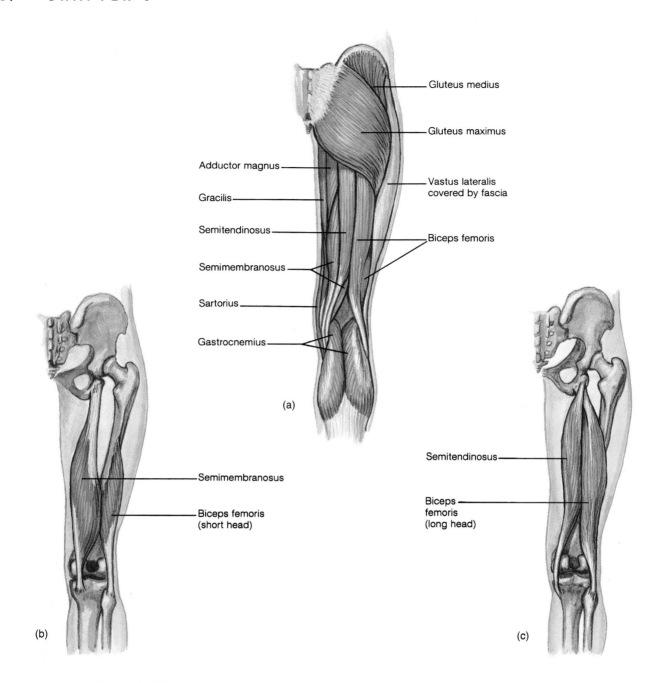

Gluteus medius

Gluteus maximus

Adductor magnus

Vastus lateralis covered by fascia

Gracilis

Semitendinosus

Biceps femoris

Semimembranosus

Sartorius

Gastrocnemius

(a)

Semimembranosus

Biceps femoris (short head)

(b)

Semitendinosus

Biceps femoris (long head)

(c)

Figure 3-20
(a) Muscles of the posterior right thigh. (b and c) Isolated views of muscles that flex the leg at the knee.

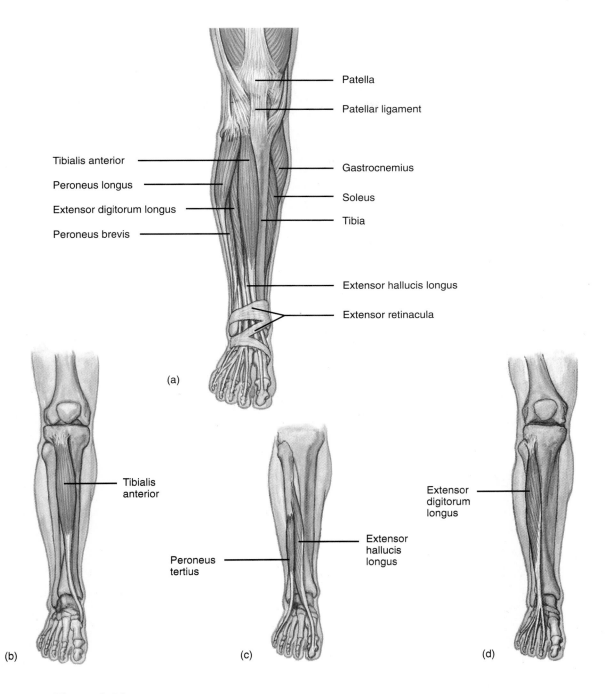

Patella

Patellar ligament

Tibialis anterior

Peroneus longus

Extensor digitorum longus

Peroneus brevis

Gastrocnemius

Soleus

Tibia

Extensor hallucis longus

Extensor retinacula

(a)

Tibialis
anterior

Peroneus
tertius

Extensor
hallucis
longus

Extensor
digitorum
longus

(b)

(c)

(d)

Figure 3-21
(a) Muscles of the anterior right leg. (b–d) Isolated views of muscles associated with the anterior leg.

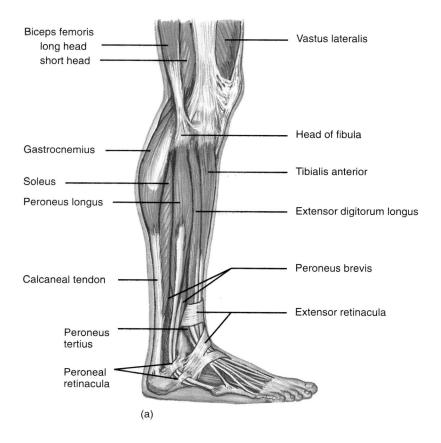

Biceps femoris
long head
short head

Vastus lateralis

Head of fibula

Gastrocnemius

Tibialis anterior

Soleus

Peroneus longus

Extensor digitorum longus

Peroneus brevis

Calcaneal tendon

Extensor retinacula

Peroneus
tertius

Peroneal
retinacula

(a)

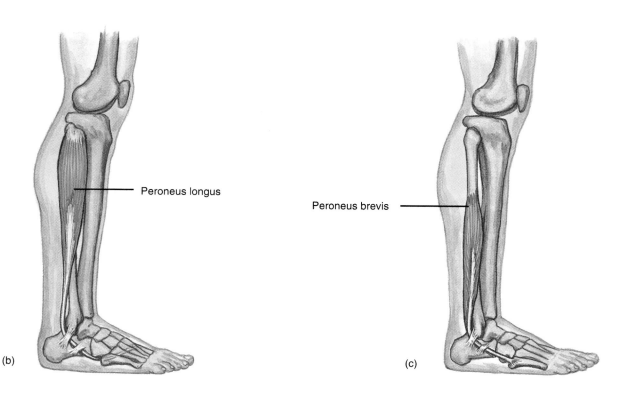

Peroneus longus

(b)

Peroneus brevis

(c)

Figure 3-22
(a) Muscles of the lateral right leg. Isolated views of (b) peroneus longus and (c) peroneus brevis.

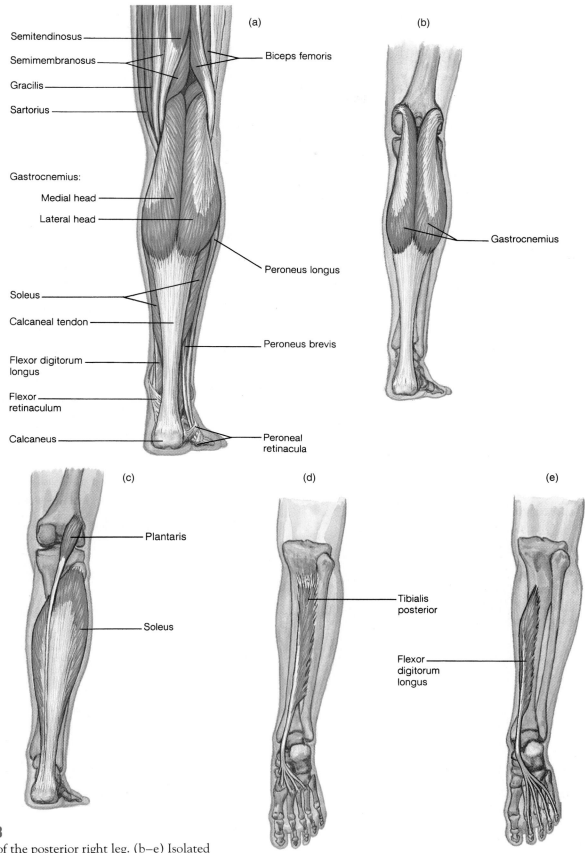

Semitendinosus

Semimembranosus

Gracilis

Sartorius

Biceps femoris

Gastrocnemius:

Medial head

Lateral head

Peroneus longus

Soleus

Calcaneal tendon

Peroneus brevis

Flexor digitorum longus

Flexor retinaculum

Calcaneus

Peroneal retinacula

Gastrocnemius

Plantaris

Soleus

Tibialis posterior

Flexor digitorum longus

(a)   (b)   (c)   (d)   (e)

Figure 3-23

(a) Muscles of the posterior right leg. (b–e) Isolated views of muscles associated with the posterior right leg.

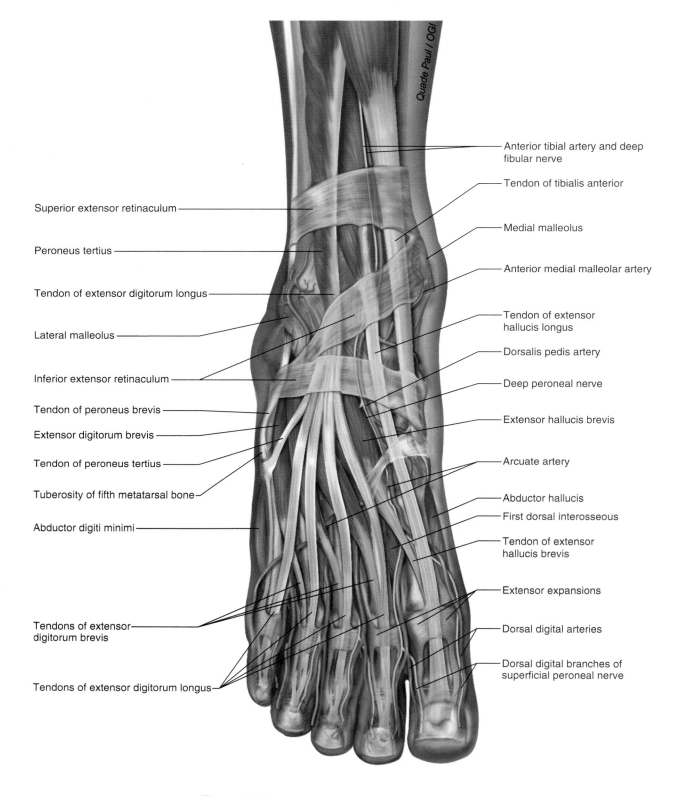

Superior extensor retinaculum

Peroneus tertius

Tendon of extensor digitorum longus

Lateral malleolus

Inferior extensor retinaculum

Tendon of peroneus brevis

Extensor digitorum brevis

Tendon of peroneus tertius

Tuberosity of fifth metatarsal bone

Abductor digiti minimi

Tendons of extensor digitorum brevis

Tendons of extensor digitorum longus

Anterior tibial artery and deep fibular nerve

Tendon of tibialis anterior

Medial malleolus

Anterior medial malleolar artery

Tendon of extensor hallucis longus

Dorsalis pedis artery

Deep peroneal nerve

Extensor hallucis brevis

Arcuate artery

Abductor hallucis

First dorsal interosseous

Tendon of extensor hallucis brevis

Extensor expansions

Dorsal digital arteries

Dorsal digital branches of superficial peroneal nerve

Figure 3-24
Dorsum of the foot, anterior view.

CHAPTER 4

Dissections

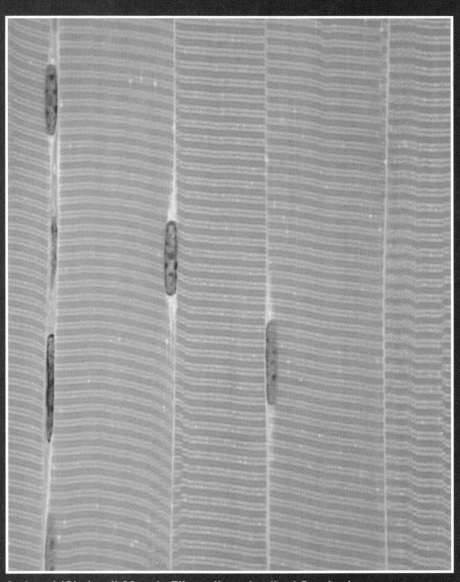

Striated (Skeletal) Muscle Fibers (Longitudinal Section)

Figure 4-1
The Cat Skeleton
1. Maxilla
2. Mandible
3. Orbit
4. Zygomatic arch
5. Cranium
6. Cervical vertebrae (7)
7. Sternum
8. Scapula
9. Humerus
10. Radius
11. Ulna
12. Carpal bones
13. Metacarpal bones
14. Phalanges
15. Thoracic vertebrae (13)
16. Ribs
17. Lumbar vertebrae (7)
18. Pelvis
19. Femur
20. Tibia
21. Fibula
22. Calcaneus
23. Tarsal bones
24. Metatarsal bones
25. Phalanges
26. Caudal vertebrae (21–25)

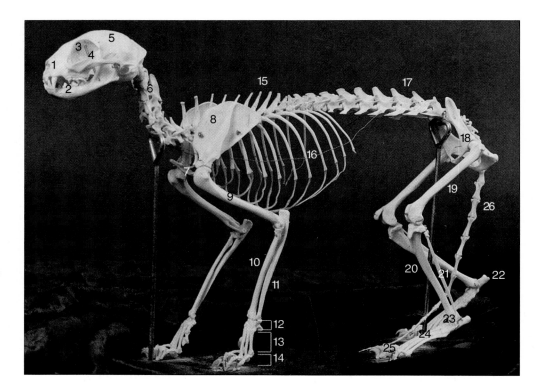

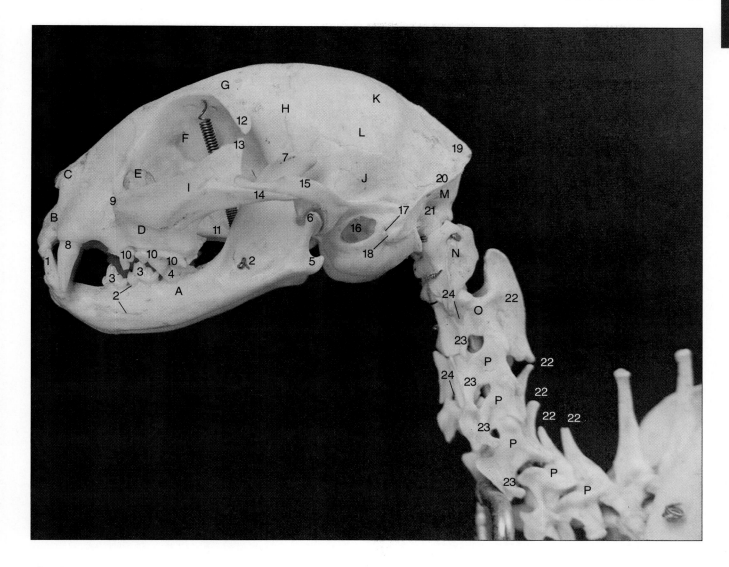

Figure 4-2
Cat Skull, Left Lateral View

A. Mandible
  1. Lower canine tooth
  2. Mental foramina
  3. Lower premolar teeth
  4. Lower molar teeth
  5. Angular process
  6. Condyloid process
  7. Coronoid process
B. Incisive bone
C. Nasal bone
D. Maxilla
  8. Upper canine tooth
  9. Infraorbital foramen
  10. Upper premolar tooth
  11. Upper molar tooth

E. Lacrimal bone and fossa
F. Orbit
G. Frontal bone
  12. Zygomatic process of frontal bone
H. Coronal suture
I. Malar or zygomatic bone
  13. Frontal process of malar
  14. Temporal process of malar
J. Temporal bone
  15. Zygomatic process of temporal bone
  16. External auditory meatus
  17. Stylomastoid foramen
  18. Mastoid process

K. Parietal bone
L. Squamosal suture
M. Occipital bone
  19. External occipital protuberance
  20. Nuchal crest
  21. Occipital condyle
N. Atlas
O. Axis
P. Cervical vertebrae (3–7)
  22. Spinous process
  23. Transverse process
  24. Transverse foramen

Figure 4-3
Axial Skeleton of Cat, Dorsal View
1. Frontal bone
2. Parietal bone
3. Sagittal suture
4. Coronal suture
5. Bregma
6. Atlas
7. Transverse process (wing) of atlas
8. Axis
9. Cervical vertebrae (7)
10. Thoracic vertebrae (13)
11. Ribs
12. Lumbar vertebrae (7)
13. Sacral vertebrae (3)
14. Caudal vertebrae (21–25)
15. Scapula
16. Humerus
17. Ilium
18. Ischium
19. Femur

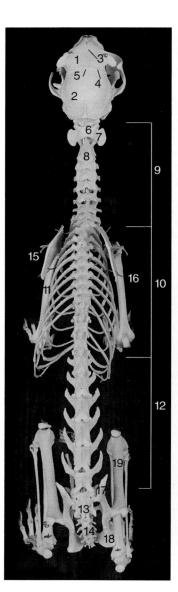

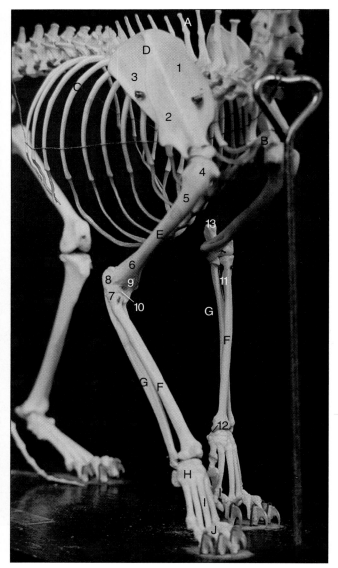

Figure 4-4
Cat Skeleton, Front Right Lateral Aspect

A. Vertebral spinous process
B. Sternum
C. Ribs
D. Scapula
  1. Supraspinous fossa
  2. Acromial spine
  3. Infraspinous fossa
E. Humerus
  4. Proximal head
  5. Deltoid tuberosity
  6. Distal head
  7. Trochlea
  8. Lateral epicondyle
  9. Medial epicondyle
  10. Radial fossa

F. Radius
  11. Radial tuberosity
  12. Styloid process
G. Ulna
  13. Olecranon process
H. Carpal bones
I. Metacarpal bones
J. Phalanges

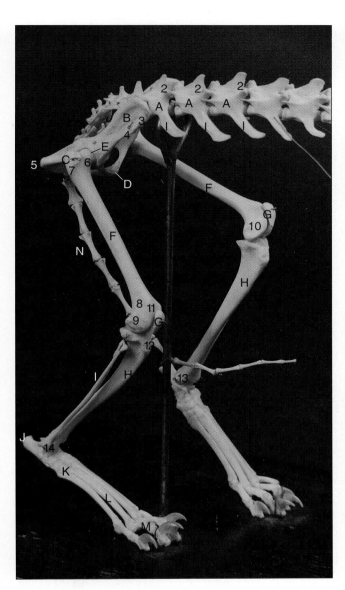

Figure 4-5
Cat Skeleton, Right Lateral Aspect
A. Lumbar vertebrae
   1. Transverse processes
   2. Spinous processes
B. Ilium
   3. Cranial ventral iliac spine
   4. Caudal ventral iliac spine
C. Ischium
   5. Ischial tuberosity
D. Pubis
E. Acetabulum
F. Femur
   6. Proximal head
   7. Greater trochanter
   8. Distal head
   9. Lateral condyle
   10. Medial condyle
   11. Trochlea
G. Patella
H. Tibia
   12. Tibial tuberosity
   13. Medial malleolus
I. Fibula
   14. Lateral malleolus
J. Calcaneus
K. Tarsal bones
L. Metatarsal bones
M. Phalanges
N. Caudal vertebrae

Figure 4-6
Superficial Anatomy of Cat Head and Neck, Left Lateral View

1. Vibrissal barrels for sensory hairs (whiskers)
2. Tongue
3. Buccinator muscle
4. Diagastric muscle
5. Temporalis muscle
6. Masseter muscle
7. Dorsal buccal branch of facial (VII) nerve
8. Ventral buccal branch of facial (VII) nerve
9. Parotid duct
10. Parotid gland
11. Submandibular gland
12. Lymph node
13. External jugular vein
14. Transverse jugular vein
15. Anterior facial vein
16. Posterior facial vein
17. Sternohyoid muscle
18. Sternothyroid muscle
19. Cleidomastoid muscle
20. Sternomastoid muscle
21. Clavotrapezius muscle
22. Clavobrachialis muscle
23. Acromiotrapezius muscle

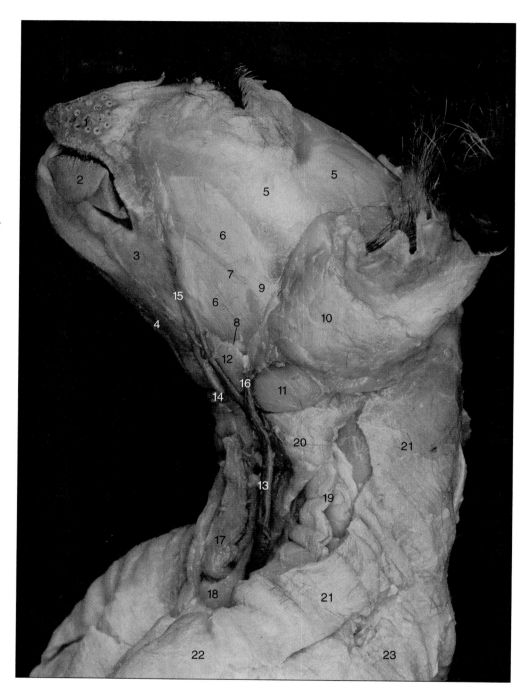

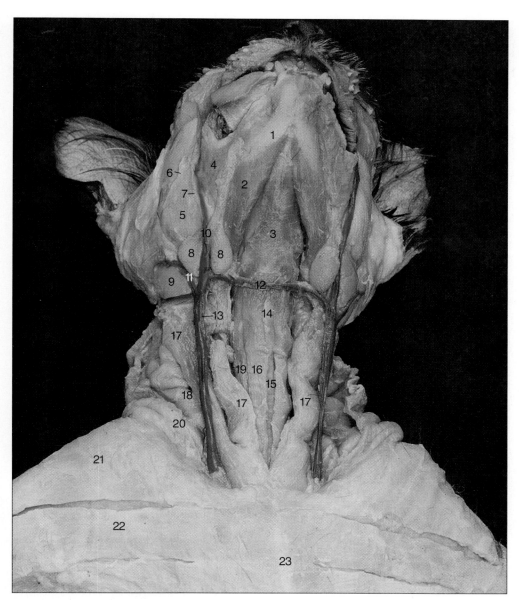

Figure 4-7
Superficial Anatomy of Cat Head and Neck, Ventral Aspect
1. Body of mandible
2. Digastric muscle
3. Mylohyoid muscle
4. Buccinator muscle
5. Masseter muscle
6. Dorsal branch of facial (VII) nerve
7. Ventral branch of facial (VII) nerve
8. Lymph node
9. Submandibular gland
10. Anterior facial vein
11. Posterior facial vein
12. Transverse jugular vein
13. External jugular vein
14. Larynx
15. Trachea
16. Sternomastoid muscle
17. Sternothyroid muscle (unavoidably damaged on animal's right side during vascular perfusion)
18. Cleidomastoid muscle
19. Sternothyroid muscle
20. Clavotrapezius muscle
21. Clavobrachialis muscle
22. Pectoantebrachialis muscle
23. Sternum

Figure 4-8
Deep Anatomy of Cat Head and Neck, Left Ventrolateral Aspect

1. Lower canine tooth
2. Upper canine tooth
3. Upper premolar tooth
4. Lower premolar tooth
5. Body of mandible
6. Digastric muscle
7. Mylohyoid muscle
8. Temporalis muscle
9. Masseter muscle
10. Dorsal branch of facial (VII) nerve
11. Ventral branch of facial (VII) nerve
12. Parotid duct
13. Cutaneous branch of facial (VII) nerve
14. Platysma muscle (reflected)
15. Lymph node
16. Sternohyoid muscle
17. Sternomastoid muscle (reflected)
18. Cleidomastoid muscle
19. Omohyoid muscle
20. 4th cervical nerve
21. 5th cervical nerve
22. Jugular vein
23. Subclavian vein
24. Musculocutaneous nerve
25. Radial nerve
26. Median nerve
27. Ulnar nerve
28. Thoracic nerve
29. Ventral thoracic nerve (cut)
30. Axillary nerve

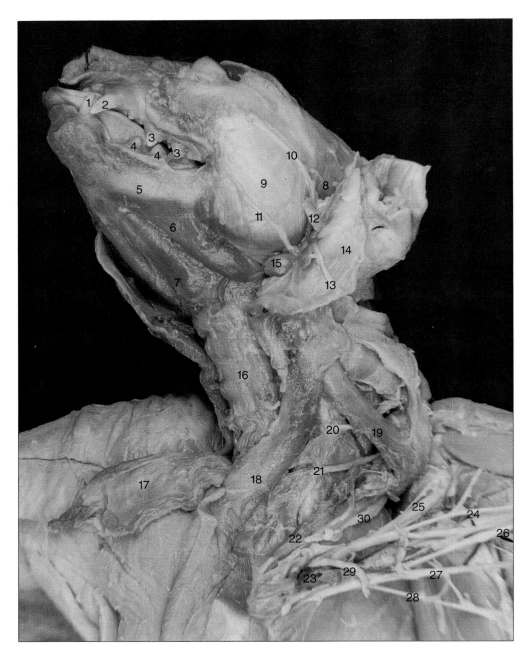

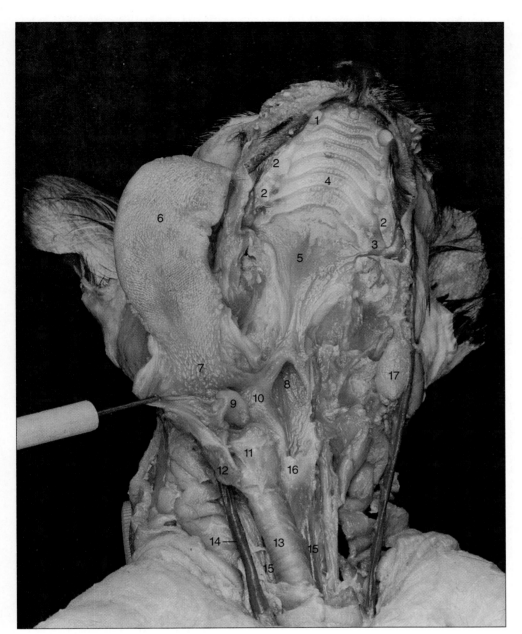

Figure 4-9
Deep Anatomy of Cat Head and Neck, Oral Cavity with Mandible Removed

1. Upper canine tooth
2. Upper premolar tooth
3. Upper molar tooth
4. Hard palate with palatine rugae
5. Soft palate
6. Tongue
7. Foliate papillae
8. Isthmus of fauces
9. Epiglottis
10. Palatine tonsil
11. Larynx
12. Thyroid gland (reflected)
13. Trachea
14. External jugular vein
15. Carotid artery
16. Esophagus
17. Lymph node

Figure 4-10
**Superficial Muscles of the Cat
Thoracic Limb, Ventral Aspect**
1. Clavobrachialis muscle
2. Pectoantebrachialis muscle
3. Pectoralis major muscle
4. Pectoralis minor muscle
5. Latissimus dorsi muscle
6. Epitrochlearis muscle
7. Flexor carpi ulnaris muscle
8. Palmaris longus muscle
9. Flexor carpi radialis muscle
10. Pronator teres muscle
11. Extensor carpi radialis
    muscle
12. Brachioradialis muscle
    (cut)
13. Cephalic vein
14. Antebrachial fascia
15. Ulnar nerve
16. Olecranon process of ulna
17. Flexor retinaculum
    (transverse carpal
    ligament)

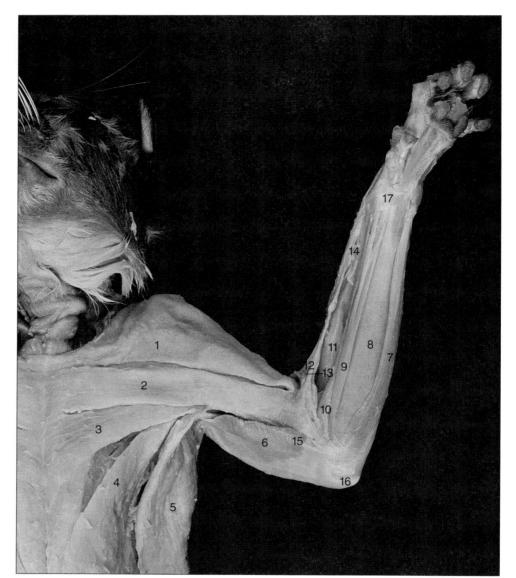

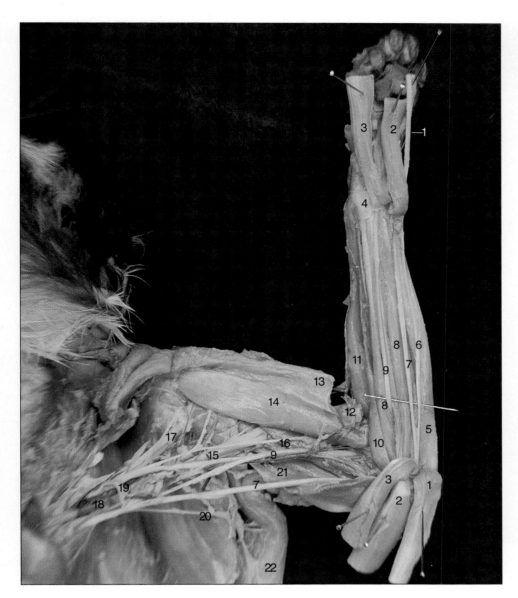

Figure 4-11
Deep Muscles of the Cat
Left Thoracic Limb,
Ventral Aspect

1. Flexor carpi ulnaris muscle (cut and reflected)
2. Palmaris longus muscle (cut and reflected)
3. Flexor carpi radialis muscle (cut and reflected)
4. Flexor retinaculum
5. Extensor carpi ulnaris
6. Cutaneous branch of ulnar nerve
7. Ulnar nerve
8. Flexor digitorum profundus
9. Median nerve
10. Pronator teres muscle
11. Extensor carpi radialis muscle
12. Brachioradialis muscle (cut)
13. Clavobrachialis muscle (cut and reflected)
14. Biceps brachii muscle
15. Radial nerve
16. Musculocutaneous nerve
17. Axillary nerve
18. Subclavian vein
19. Ventral thoracic nerve (cut) ·
20. Thoracic nerve
21. Triceps brachii muscle
22. Latissimus dorsi muscle

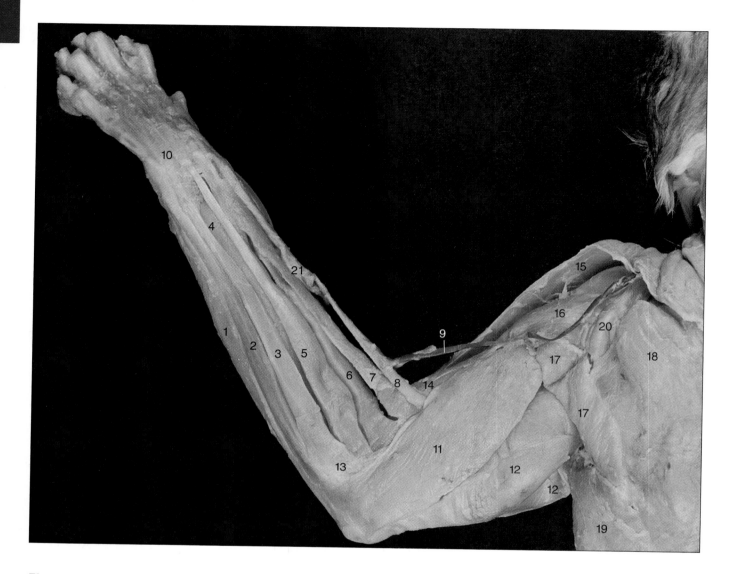

Figure 4-12
Superficial Muscles of the Cat Left Thoracic Limb, Dorsal Aspect

1. Flexor carpi ulnaris muscle
2. Extensor carpi ulnaris muscle
3. Extensor carpi digitorum lateralis muscle
4. Extensor pollicis brevis muscle
5. Extensor digitorum communis muscle
6. Extensor carpi radialis brevis muscle
7. Extensor carpi radialis longus muscle
8. Brachioradialis muscle
9. Cephalic vein
10. Extensor retinaculum (dorsal carpal ligament)
11. Triceps brachii muscle (lateral head)
12. Triceps brachii muscle (long head)
13. Anconeus muscle
14. Brachialis muscle
15. Clavobrachialis muscle
16. Acromiodeltoid muscle
17. Spinodeltoid muscle
18. Acromiotrapezius muscle
19. Latissimus dorsi muscle
20. Levator scapulae ventralis
21. Antebrachial fascia

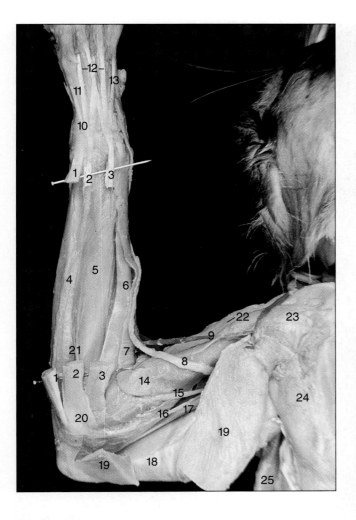

Figure 4-13
Deep Muscles of the Cat Left Thoracic Limb, Dorsal Aspect

1. Extensor carpi ulnaris muscle (cut)
2. Extensor digitorum lateralis muscle (cut)
3. Extensor digitorum communis muscle (cut)
4. Extensor indicis proprius muscle
5. Extensor pollicis brevis muscle
6. Extensor carpi radialis muscle
7. Brachioradialis muscle
8. Radial nerve
9. Cephalic vein
10. Extensor retinaculum (dorsal carpal ligament)
11. Extensor digiti minimi tendon
12. Extensor digitorum tendons
13. Extensor indicis tendon
14. Brachioradialis muscle
15. Median nerve
16. Ulnar nerve
17. Triceps brachii muscle (medial head)
18. Triceps brachii muscle (long head)
19. Triceps brachii muscle (lateral head, cut)
20. Anconeus muscle
21. Posterior interosseous nerve
22. Clavobrachialis muscle
23. Acromiodeltoid muscle
24. Spinodeltoid muscle
25. Latissimus dorsi muscle

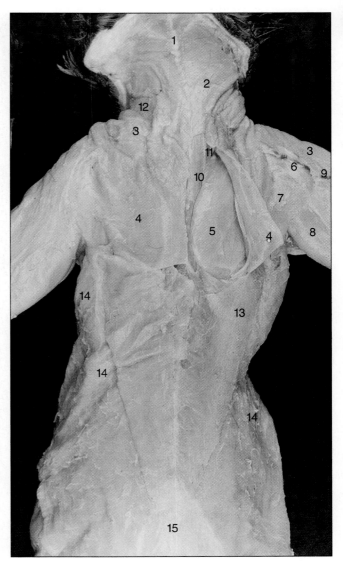

Figure 4-14
Superficial Muscles of the Cat Neck and Back

1. Nuchal ligament
2. Clavotrapezius muscle
3. Clavobrachialis muscle
4. Acromiotrapezius muscle (cut and reflected on right)
5. Supraspinatus muscle
6. Acromiodeltoid muscle
7. Spinodeltoid muscle
8. Triceps brachii muscle (long head)
9. Cephalic vein
10. Rhomboideus minor muscle
11. Rhomboideus capitis muscle (occipito-scapularis muscle)
12. Splenius capitis muscle
13. Spinotrapezius muscle
14. Latissimus dorsi muscle
15. Lumbodorsal fascia

Figure 4-15
Deep Muscles of the Cat Neck and Back

1. Nuchal ligament
2. Clavotrapezius muscle (reflected on left)
3. Acromiotrapezius muscle (cut, removed altogether on left)
4. Supraspinatus muscle
5. Infraspinatus muscle
6. Triceps brachii muscle (long head)
7. Triceps brachii muscle (lateral head)
8. Acromiodeltoid muscle
9. Clavobrachialis muscle
10. Rhomboideus capitis muscle
11. Splenius capitis muscle
12. Rhomboideus minor muscle
13. Rhomboideus major muscle
14. Spinotrapezius muscle
15. Latissimus dorsi muscle (reflected on left, partially removed on right)
16. Multifidus muscle
17. Spinalis muscle
18. Longissimus muscle
19. Iliocostalis muscle
20. Lumbodorsal fascia (largely removed)

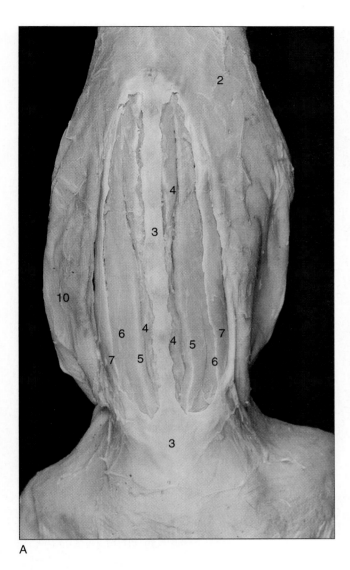

A

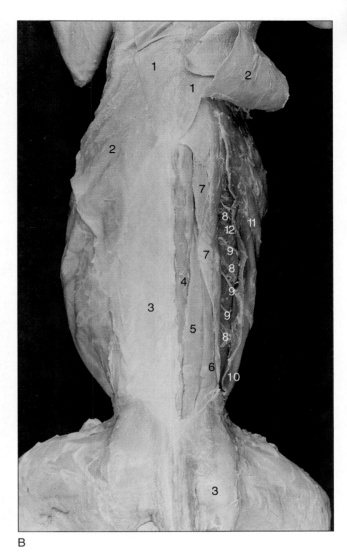

B

Figure 4-16
Deep Muscles of the Back of the Cat

1. Spinotrapezius muscle
2. Latissimus dorsi muscle (cut and rolled on right)
3. Lumbodorsal fascia
4. Multifidus muscle
5. Spinalis muscle
6. Longissimus muscle
7. Iliocostalis muscle
8. Rib
9. Dorsal ramus of spinal nerve
10. External oblique muscle
11. External intercostal muscle
12. Internal intercostal muscle

Figure 4-17
**Superficial Muscles of
the Cat Left Hind Limb,
Dorsal Aspect**

1. Lumbodorsal fascia
2. Sartorius muscle
3. Tensor fascia latae
   muscle
4. Iliotibial band
5. Gluteus medius muscle
6. Gluteus maximus muscle
7. Caudofemoralis muscle
8. Biceps femoris muscle
9. Semitendinosus muscle
10. Semimembranosus
    muscle
11. Gastrocnemius muscle
12. Soleus muscle
13. Achilles tendon
14. Calcaneal tuberosity
15. Flexor hallucis longus
    muscle
16. Peroneus brevis muscle
17. Peroneus longus muscle
    and tendon
18. Tibialis anterior muscle

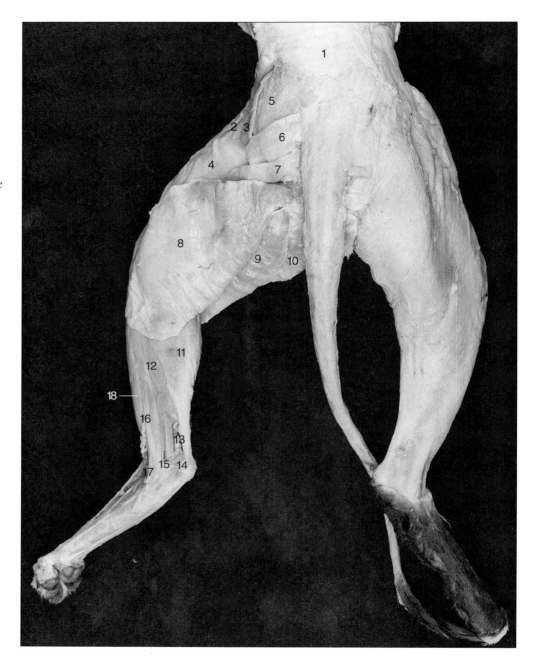

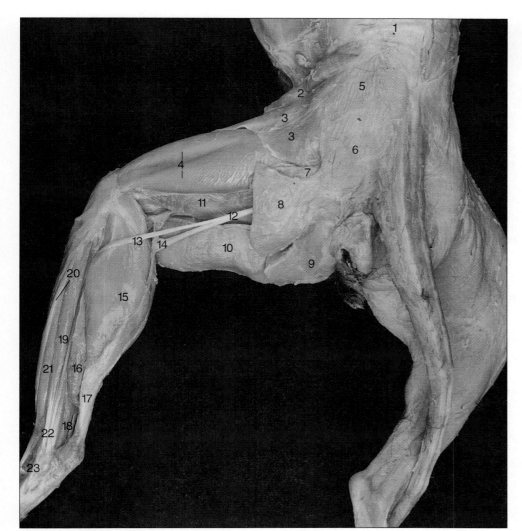

Figure 4-18
Deep Muscles of the Cat Left Hind Limb, Dorsal Aspect
 1. Lumbodorsal fascia
 2. Sartorius muscle
 3. Tensor fascia latae muscle
 4. Vastus lateralis muscle
 5. Gluteus medius muscle (under fascia)
 6. Gluteus maximus muscle (under fascia)
 7. Caudofemoralis muscle
 8. Biceps femoris muscle (cut)
 9. Semitendinosus muscle (cut)
10. Semimembranosus muscle
11. Adductor femoris muscle
12. Sciatic nerve
13. Common peroneal division of sciatic nerve
14. Tibial division of sciatic nerve
15. Gastrocnemius muscle
16. Soleus muscle
17. Achilles tendon
18. Flexor hallucis longus muscle
19. Peroneus longus muscle
20. Tibialis anterior muscle
21. Extensor digitorum longus muscle
22. Proximal extensor retinaculum
23. Distal extensor retinaculum

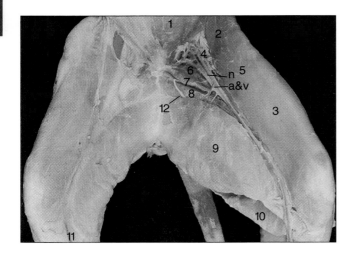
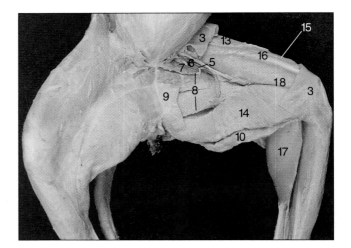

Figure 4-19
Superficial Muscles of the Cat Left Hind Limb, Medial Aspect
1. Rectus abdominis muscle
2. External oblique muscle
3. Sartorius muscle (cut)
4. Iliopsoas muscle (deep to blood vessels)
5. Femoral artery (a), vein (v), and nerve (n)
6. Pectineus muscle (deep to blood vessels)
7. Adductor longus muscle
8. Adductor femoris muscle
9. Gracilis muscle
10. Semitendinosus muscle
11. Greater saphenous vein
12. Branch of obturator nerve
13. Tensor fascia latae muscle
14. Semimembranosus muscle
15. Vastus lateralis muscle
16. Rectus femoris muscle
17. Gastrocnemius muscle
18. Vastus medialis muscle

Figure 4-20
Deep Muscles of the Cat Left Hind Limb, Medial Aspect
1. Sartorius muscle (cut)
2. Tensor fascia latae muscle
3. Vastus lateralis muscle
4. Rectus femoris muscle
5. Femoral artery (a), vein (v), and nerve (n)
6. Middle caudal femoral artery and vein
7. Pectineus muscle
8. Adductor longus muscle
9. Adductor femoris muscle
10. Gracilis muscle (cut)
11. Semimembranosus muscle
12. Semitendinosus muscle
13. Biceps femoris muscle
14. Gastrocnemius muscle (reflected)
15. Soleus muscle
16. Achilles tendon
17. Posterior tibial nerve
18. Flexor hallucis longus muscle
19. Flexor digitorum longus muscle
20. Tibialis posterior muscle
21. Tibia
22. Tibialis anterior muscle
23. Proximal extensor retinaculum
24. Vastus medialis muscle

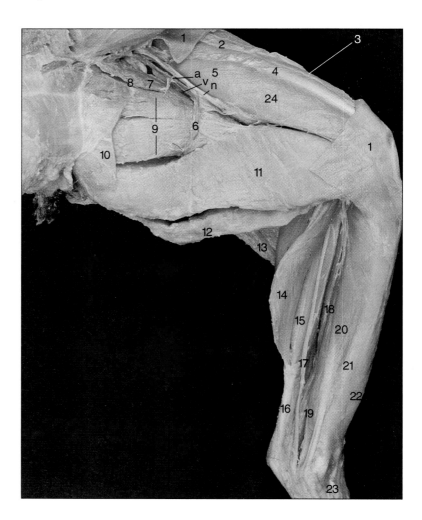

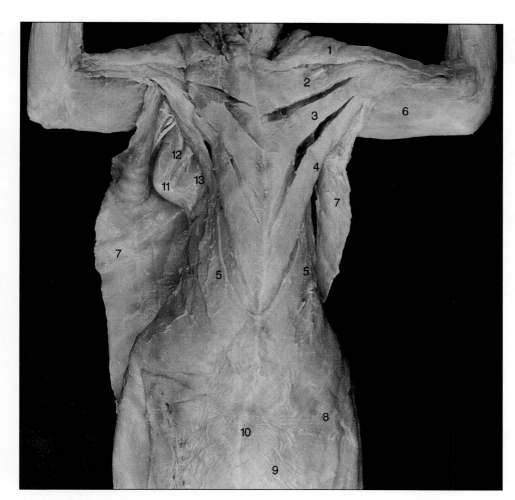

Figure 4-21
Superficial Muscles of the
Cat Thorax, Ventral View
1. Clavobrachialis muscle
2. Pectoantebrachialis muscle
3. Pectoralis major muscle
4. Pectoralis minor muscle
5. Xiphihumeralis muscle
6. Epitrochlearis muscle
7. Latissimus dorsi muscle
8. External oblique muscle
9. Rectus abdominis muscle (deep to aponeurosis)
10. Linea alba
11. Inferior angle of scapula
12. Teres major muscle
13. Subscapularis muscle

Figure 4-22
Deep Muscles of the Cat Shoulder and Thorax, Right Ventral View

1. Latissimus dorsi muscle (reflected)
2. Scalenus muscles
   a. Anterior (continuous with transversus costarum)
   b. Medius
   c. Posterior
3. Axillary artery (a) and vein (v)
4. Radial nerve
5. External jugular vein
6. Internal jugular vein
7. Thoracodorsal nerve
8. Long thoracic nerve
9. Thoracoacromial blood vessels
10. Serratus ventralis muscle
11. Teres major muscle
12. Subscapularis muscle
13. Sternum
14. Ventral thoracic nerve

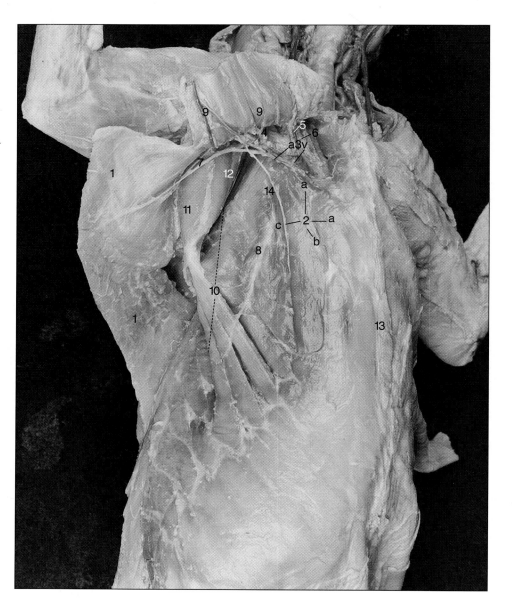

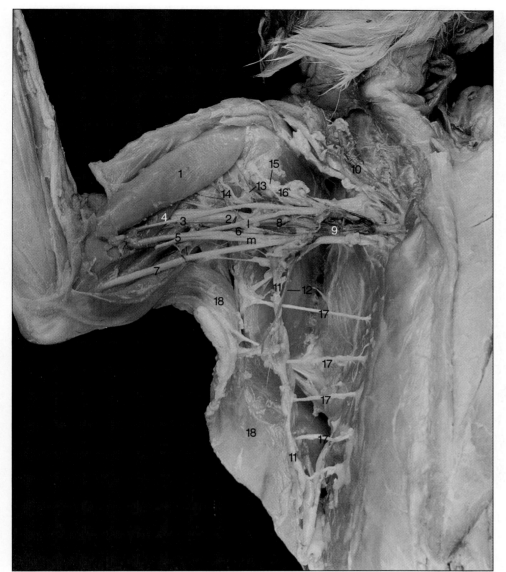

Figure 4-23
Brachial Plexus of the Cat,
Right Ventral Aspect

1. Biceps brachii muscle
2. Radial nerve
3. Musculocutaneous nerve
4. Coracobrachialis muscle
5. Median nerve
6. Lateral (l) and median (m) roots of the median nerve
7. Ulnar nerve
8. Axillary artery
9. Axillary vein
10. External jugular vein
11. Thoracodorsal nerve
12. Thoracodorsal artery
13. Thoracoacromial artery
14. Anterior circumflex humeral artery and axillary nerve
15. Caudal subscapular nerve
16. Proximal subscapular nerve
17. Dorsal rami of thoracic nerves
18. Latissimus dorsi muscle (reflected)

Figure 4-24
Thoracic Cavity of the Cat
1. Heart within
   pericardium
2. Thymus gland
3. Diaphragm
4. Lung, anterior lobe
5. Lung, middle lobe
6. Lung, posterior lobe
7. Ribs (cut)

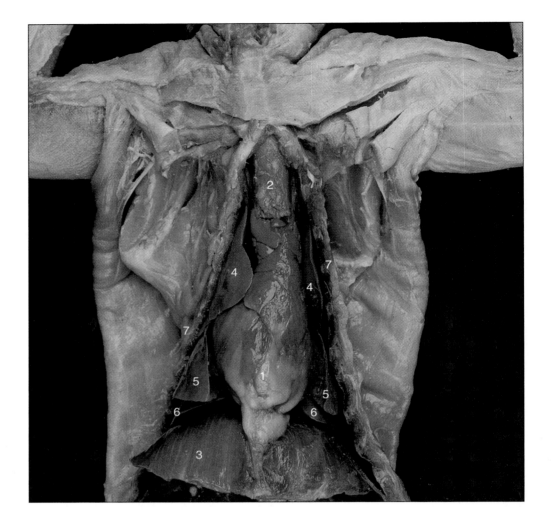

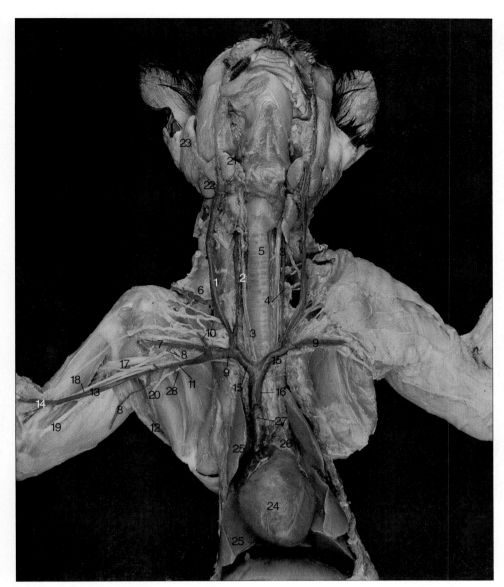

Figure 4-25
Major Veins of the Cat,
Neck and Thorax

1. External jugular vein
2. Internal jugular vein
3. Common carotid artery (right)
4. Vagus nerve
5. Trachea
6. Transverse scapular vein
7. Subscapular vein
8. Thoracodorsal vein
9. Subclavian vein
10. Cephalic vein
11. Axillary vein
12. Latissimus dorsi muscle
13. Brachial vein
14. Median cubital vein
15. Innominate (brachiocephalic) vein
16. Anterior vena cava
17. Radial nerve
18. Median nerve
19. Ulnar nerve
20. Thoracodorsal nerve
21. Lymph node
22. Submaxillary gland
23. Parotid gland
24. Heart
25. Lung
26. Thymus gland
27. Anterior thoracic vein (cut) (internal mammary vein)
28. Long thoracic vein (cut)

Figure 4-26
Major Arteries of the Cat, Neck and Thorax

1. Common carotid artery
2. Vagus nerve
3. Vertebral artery
4. Transverse scapular artery
5. Axillary artery
6. Brachial artery
7. Thoraco-acromial artery (a) and nerve (n)
8. Musculocutaneous nerve
9. Median nerve
10. Radial nerve
11. Ulnar nerve
12. Trachea
13. Esophagus (displaced to animal's left from normal position posterior to trachea)
14. Phrenic nerve
15. Right subclavian artery
16. Innominate (brachiocephalic) artery
17. Left subclavian artery
18. Aortic arch
19. Anterior vena cava (cut)
20. Teres major muscle
21. Subscapularis muscle
22. Biceps brachii muscle
23. Triceps brachii muscle (long head)
24. Heart
25. Lung, anterior lobe
26. Lung, middle lobe
27. Lung, mediastinal lobe
28. Lung, posterior lobe
29. Right auricle
30. Diaphragm

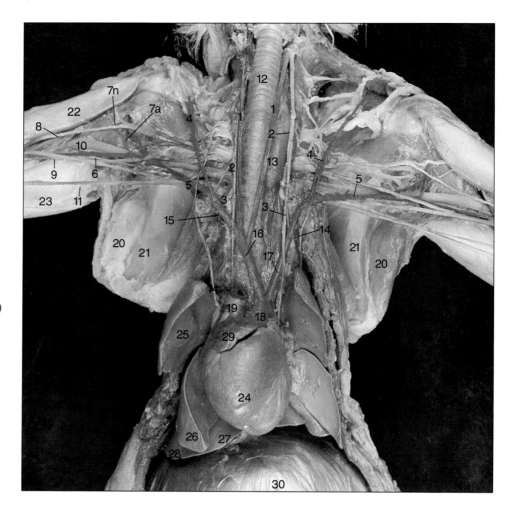

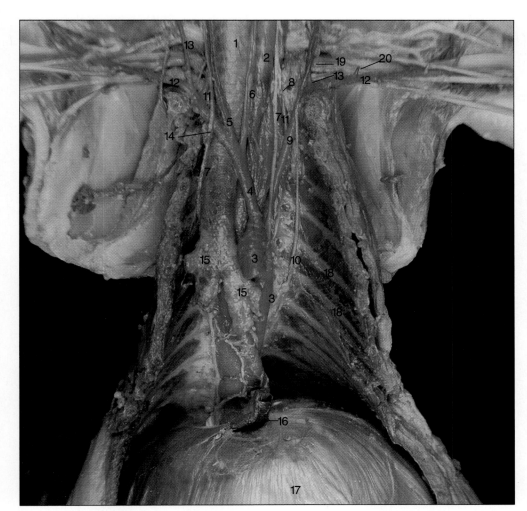

Figure 4-27
Thorax of the Cat, Heart and Lungs Removed
1. Trachea
2. Esophagus
3. Aortic arch
4. Brachiocephalic artery
5. Right common carotid artery
6. Left common carotid artery
7. Vagus nerve
8. Sympathetic trunk
9. Left subclavian artery
10. Phrenic nerve
11. Vertebral artery
12. Subclavian artery
13. Thyrocervical artery
14. Internal mammary artery
15. Right and left primary bronchi
16. Inferior vena cava
17. Diaphragm
18. Rib
19. Transverse scapular artery
20. Subscapular artery (cut)

A

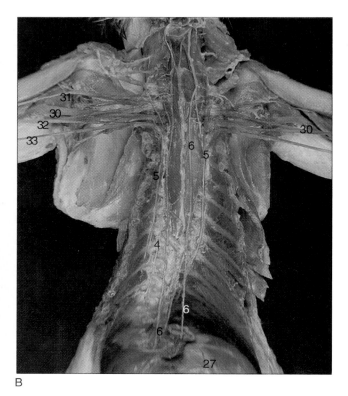

B

Figure 4-28
Veins, Arteries, and Nerves of the Cat Neck and Thorax
A, Veins Removed on Cat's Left Side, Heart Reflected to Right
B, Arteries, Veins, and Thoracic Viscera Removed

1. Larynx
2. Thyroid gland (reflected)
3. Common carotid artery
4. Vagus nerve
5. Sympathetic trunk (In **A**, two pins have been placed along sympathetic trunk. Upper pin head is just caudal to swelling of sympathetic trunk, the superior cervical ganglion. Lower transverse pin is just proximal to similar swelling, the middle cervical ganglion.)
6. Phrenic nerve
7. Aorta

8. Spinal accessory nerve (XI)
9. Spinal nerves IV, V, and VI
10. Brachial plexus
11. Lymph node (reflected and pinned)
12. Soft palate (cut)
13. Eustachian tubes (hidden behind reflected tissue of soft palate)
14. Epiglottis
15. Internal jugular vein
16. External jugular vein
17. Subscapular vein
18. Brachial vein
19. Axillary vein
20. Subclavian vein

21. Innominate (brachiocephalic) vein
22. Anterior vena cava
23. Azygous vein (cut)
24. Heart (reflected to cat's right)
25. Right auricle
26. Left auricle
27. Diaphragm
28. Esophagus
29. Trachea
30. Radial nerve
31. Musculocutaneous nerve
32. Median nerve
33. Ulnar nerve
34. Caudal subscapular nerve
35. Axillary nerve

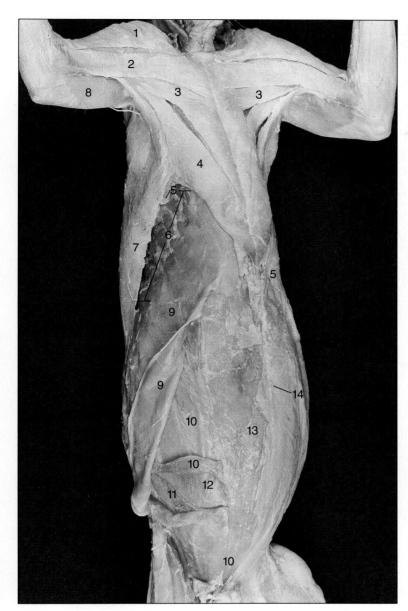

Figure 4-29
Superficial Muscles of the Cat, Abdomen and Thorax
1. Clavobrachialis muscle
2. Pectoantebrachialis muscle
3. Pectoralis major muscle
4. Pectoralis minor muscle
5. Xiphihumeralis muscle (removed on right)
6. Serratus anterior muscle
7. Latissimus dorsi muscle (cut to reveal underlying muscles)
8. Epitrochlearis muscle
9. External oblique muscle (partially reflected)
10. Internal oblique muscle (partially reflected)
11. Transversalis abdominis muscle
12. Peritoneum
13. Rectus abdominis muscle
14. Linea alba

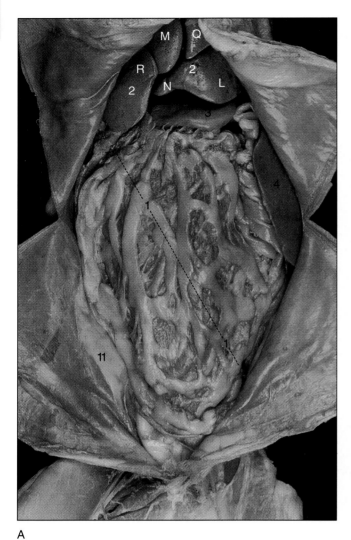

A

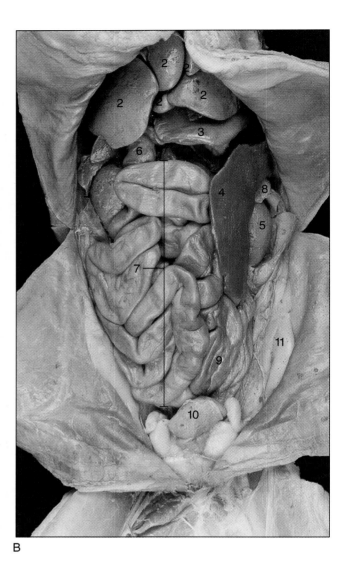

B

Figure 4-30
Abdominal Viscera of Cat with (A) Greater Omentum Intact and (B) Greater Omentum Removed

1. Greater omentum
2. Lobes of the liver
   R. Right lateral lobe
   M. Right medial lobe
   Q. Quadrate lobe
   N. Left medial lobe
   L. Left lateral lobe
3. Stomach (greater curvature)
4. Spleen
5. Kidney
6. Small intestine (duodenum)
7. Small intestine (jejuneum and ileum)
8. Pancreas
9. Large intestine (descending colon)
10. Urinary bladder
11. Abdominal fat

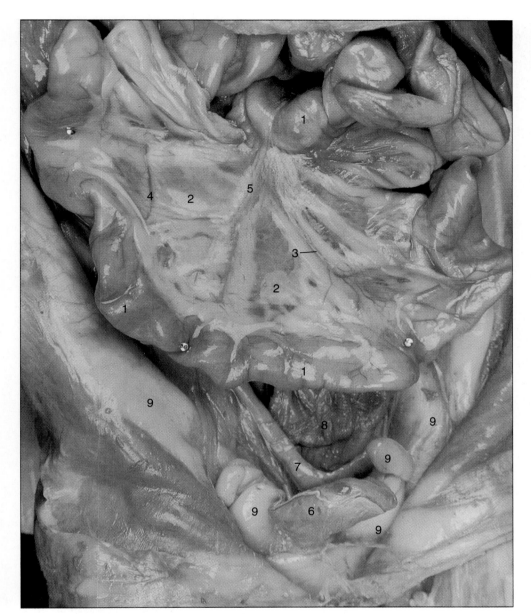

Figure 4-31
Abdominal Viscera of the Cat, Mesentery
1. Small intestine
2. Mesentery
3. Mesenteric artery
4. Mesenteric vein
5. Lymph vessel
6. Urinary bladder
7. Uterus
8. Rectum
9. Abdominal fat

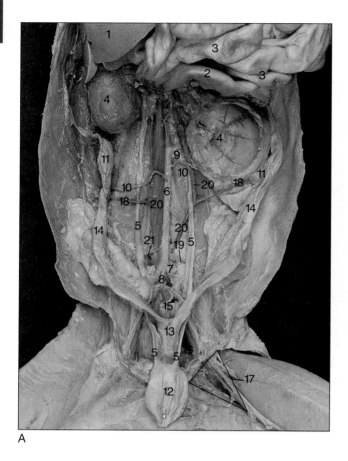
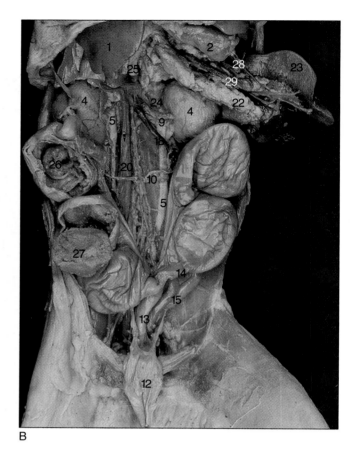

A

B

Figure 4-32
Urogenital System of the Female Cat (A) Nonpregnant and (B) Pregnant

1. Liver
2. Stomach (pylorus)
3. Small intestine
4. Kidney
5. Ureter
6. Abdominal aorta
7. External iliac artery
8. Caudal (median sacral) artery
9. Renal artery
10. Ovarian artery
11. Ovary
12. Urinary bladder (reflected and pinned)
13. Uterus
14. Uterine horn (in **B**, left horn contains two fetuses; right horn, three fetuses)
15. Rectum (cut in **A**)
16. External iliac artery and vein
17. Femoral triangle (containing femoral nerve, artery, and vein)
18. Ovarian vein
19. Iliolumbar artery and vein
20. Abdominal vena cava (split into two parallel vessels in **A**)
21. Iliolumbar artery
22. Pancreas
23. Spleen
24. Adrenal gland
25. Hepatic portal vein (cut)
26. Fetus
27. Placenta
28. Left gastroepiploic vein
29. Right gastroepiploic vein

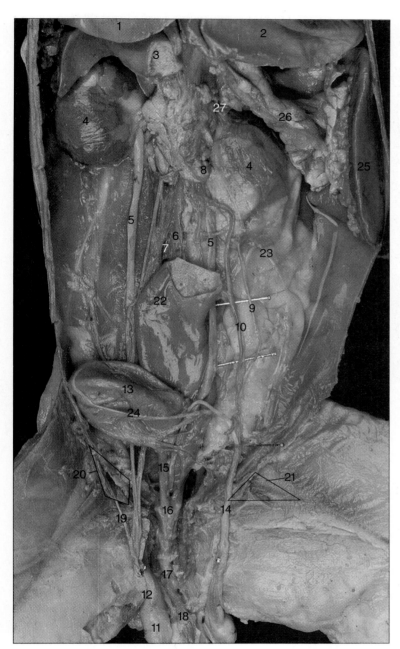

Figure 4-33
Urogenital System of the Male Cat

1. Liver
2. Stomach
3. Small intestine (duodenum, cut)
4. Kidney
5. Ureter
6. Abdominal aorta
7. Abdominal vena cava
8. Renal artery
9. Internal spermatic artery
10. Spermatic vein
11. Testis
12. Epididymis
13. Urinary bladder (reflected)
14. Vas deferens in spermatic cord *or ductus deferens*
15. Urethra
16. Prostate gland
17. Bulbourethral (Cowper's) gland
18. Penis
19. Ligament of cremaster muscle
20. External inguinal ring
21. Femoral triangle
22. Rectum (cut)
23. Lumbar nerve (medial branch)
24. Umbilical (allantoic) artery
25. Spleen
26. Pancreas
27. Adrenal gland

spermatic cord
hold ductus deferace
as well as vessals
+ nerves

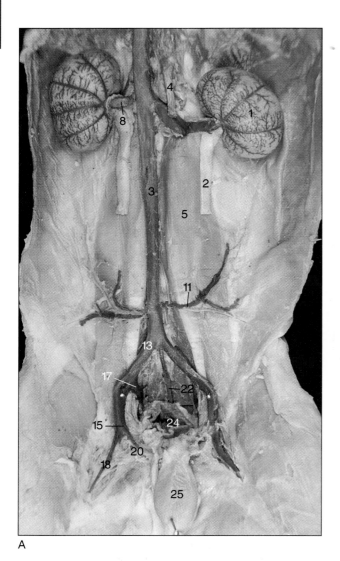

A

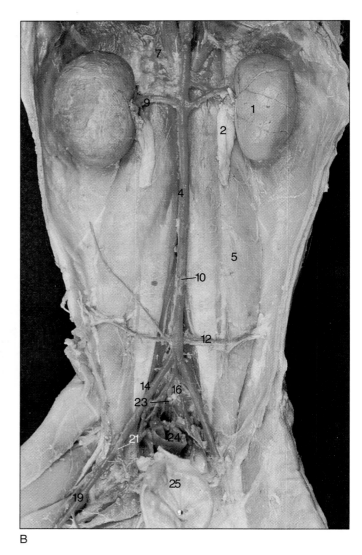

B

Figure 4-34
Major (A) Veins and (B) Arteries of the Cat Abdominopelvic Wall

1. Kidney
2. Ureter (cut and largely removed)
3. Abdominal vena cava (removed in **B**)
4. Abdominal aorta (cut and removed in **A**)
5. Psoas muscle
6. Celiac artery (cut and removed)
7. Superior mesenteric artery (cut and removed)
8. Renal vein
9. Renal artery
10. Inferior mesenteric artery (cut and removed)
11. Iliolumbar vein
12. Iliolumbar artery
13. Common iliac vein
14. External iliac artery (no common iliac artery in cats)
15. External iliac vein
16. Internal iliac (hypogastric) artery
17. Internal iliac (hypogastric) vein
18. Femoral vein
19. Femoral artery
20. Deep femoral vein
21. Deep femoral artery
22. Caudal vein
23. Median sacral (caudal) artery
24. Rectum (cut)
25. Urinary bladder (reflected and pinned)

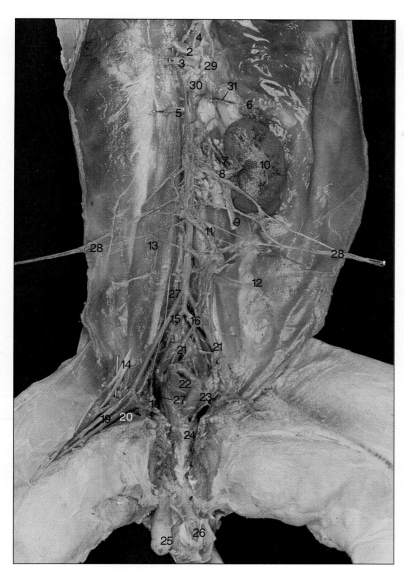

Figure 4-35
Nerves and Vessels of the Posterior Abdominopelvic Wall of the Cat

1. Abdominal aorta
2. Celiac artery (pinned)
3. Superior mesenteric artery (pinned)
4. Crus of diaphragm
5. Right adrenolumbar (phrenicoabdominal) artery (cut)
6. Adrenal gland
7. Renal vein
8. Renal artery
9. Ureter (cut)
10. Kidney
11. Inferior mesenteric artery
12. Iliolumbar artery
13. Psoas muscle
14. Femoral nerve
15. External iliac artery
16. Internal iliac artery
17. External iliac vein
18. Deep femoral artery and vein
19. Femoral artery
20. Femoral vein
21. Spermatic artery
22. Rectum (cut)
23. Urethra (cut and urinary bladder removed)
24. Prostate gland
25. Testis
26. Penis
27. Genitofemoral nerve
28. Distribution of sympathetic trunk (pinned out bilaterally)
29. Celiac ganglion
30. Superior mesenteric ganglion
31. Left adrenolumbar artery and vein

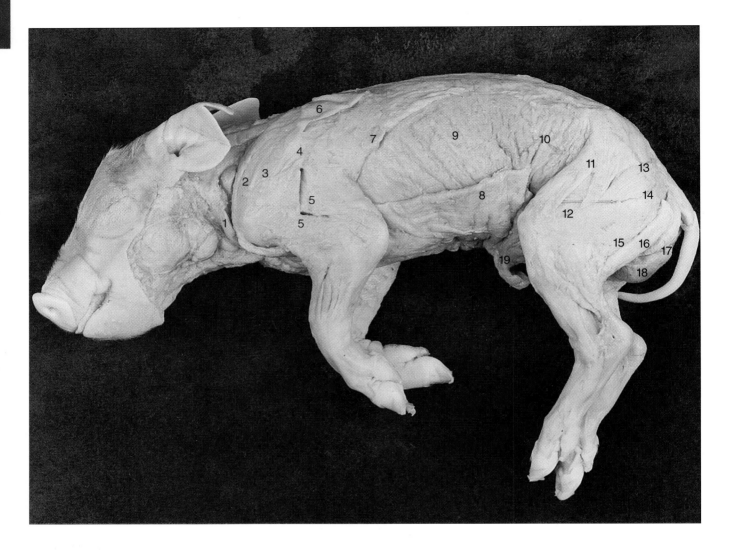

Figure 4-36
Superficial Muscles of the Fetal Pig, Left Lateral View

1. Clavotrapezius muscle
2. Clavobrachialis muscle
3. Acromiodeltoid muscle
4. Spinodeltoid muscle
5. Triceps brachii muscle
6. Spinotrapezius muscle (cut)
7. Latissimus dorsi muscle

8. External oblique muscle (cut)
9. Serratus anterior muscle
10. Internal oblique muscle
11. Tensor fascia latae muscle (split)
12. Vastus lateral muscle (under pin)
13. Gluteus medius muscle
14. Gluteus maximus muscle

15. Biceps femoris muscle
16. Semitendinosus muscle
17. Semimembranosus muscle
18. Testis
19. Umbilical cord

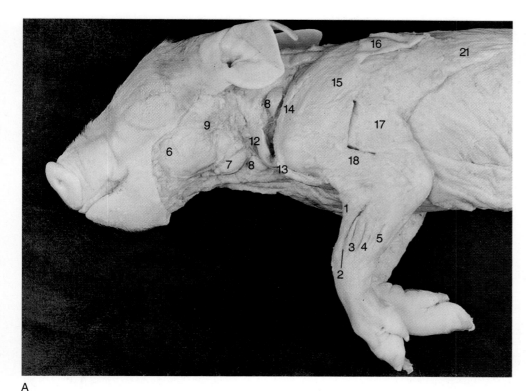

A

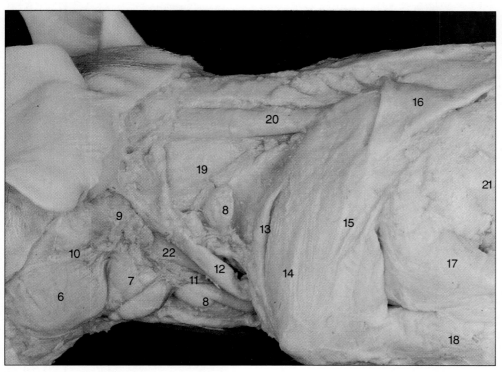

B

Figure 4-37
**Superficial Structures of the
Neck, Shoulder, and Thoracic
Limb of the Fetal Pig,
Left Lateral View**

1. Brachioradialis muscle
2. Extensor carpi radialis
   muscle
3. Extensor digitorum
   communis muscle
4. Extensor digitorum
   lateralis muscle
5. Extensor carpi ulnaris
   muscle
6. Masseter muscle
7. Submaxillary gland
8. Lymph node
9. Parotid gland
10. Salivary duct
11. External jugular vein
12. Clavotrapezius muscle
13. Clavobrachialis muscle
14. Acromiodeltoid muscle
15. Spinodeltoid muscle
16. Spinotrapezius muscle
    (cut)
17. Triceps brachii muscle
    (long head)
18. Triceps brachii muscle
    (lateral head)
19. Splenius capitis muscle
20. Rhomboideus capitis
    muscle
21. Latissimus dorsi muscle
22. Sternomastoid muscle

Figure 4-38
**Superficial Muscles of the
Hind Limb of the Fetal Pig,
Left Lateral View**
  1. Lumbodorsal fascia
  2. External oblique muscle
     (reflected)
  3. Internal oblique muscle
  4. Tensor fascia latae
     muscle (split)
  5. Vastus lateralis muscle
     (under pin)
  6. Gluteus medius muscle
  7. Gluteus maximus muscle
  8. Biceps femoris muscle
  9. Semitendinosus muscle
 10. Semimembranosus
     muscle
 11. Testis
 12. Gastrocnemius muscle
 13. Soleus muscle
 14. Achilles tendon
 15. Flexor hallucis longus
     muscle
 16. Tibialis anterior muscle

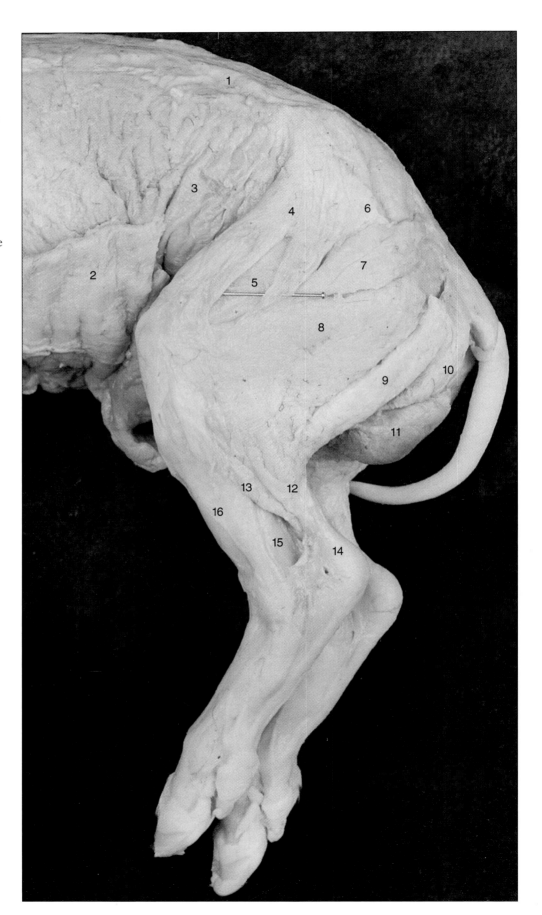

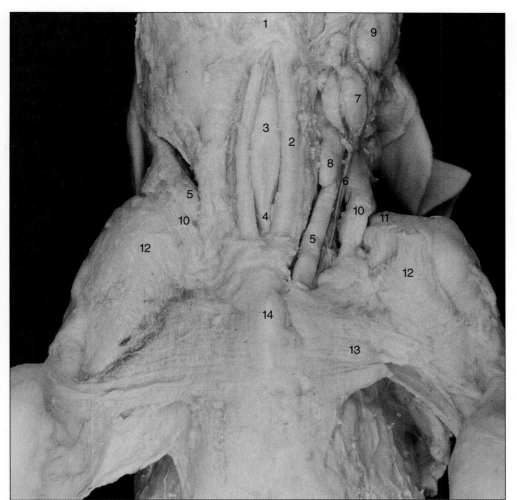

Figure 4-39
Superficial Anatomy of the Fetal Pig Neck and Shoulders, Ventral Aspect
1. Mylohyoid muscle
2. Sternohyoid muscle
3. Larynx
4. Trachea
5. Sternomastoid muscle
6. External jugular vein
7. Lymph node
8. Submaxillary gland
9. Masseter muscle
10. Clavotrapezius muscle
11. Acromiodeltoid muscle
12. Clavobrachialis muscle
13. Pectoralis major muscle
14. Sternum

Figure 4-40
Deep Anatomy of the Fetal Pig, Neck and Thorax

1. Larynx
2. Trachea
3. Thyroid gland
4. Common carotid artery
5. Vagus nerve
6. Internal jugular vein
7. External jugular vein (pinned bilaterally)
8. Cephalic vein
9. Subclavian vein
10. Superior vena cava
11. Internal mammary vein (cut, laid on lung tissue)
12. Right auricle
13. Left auricle
14. Heart
15. Lung
16. Diaphragm

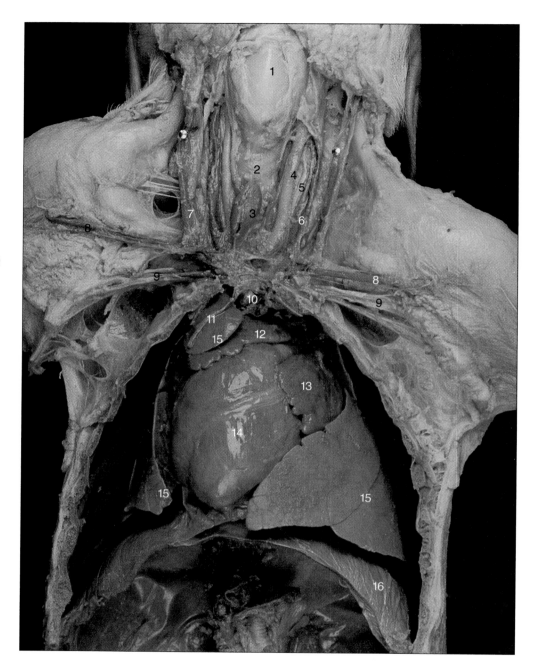

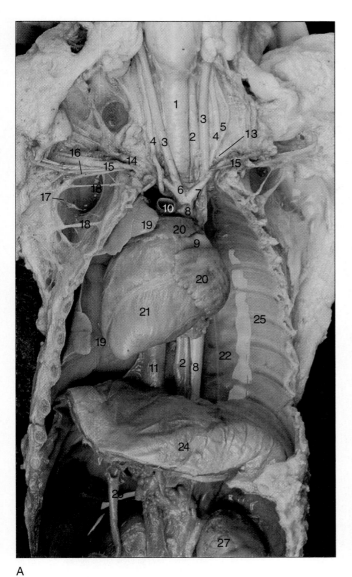

A

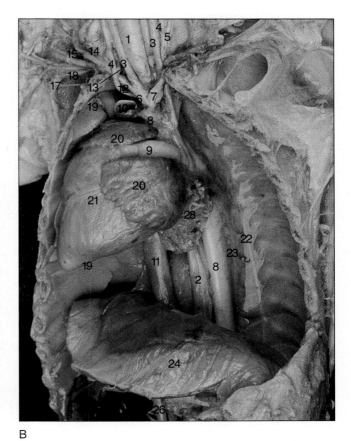

B

Figure 4-41
Arteries of the Neck and Thorax of the Fetal Pig, Left Lung Removed
A, Heart in Normal Position
B, Heart Reflected to Right

1. Trachea
2. Esophagus
3. Common carotid artery
4. Vagus nerve
5. Sympathetic trunk
6. Right innominate (brachiocephalic) artery
7. Left innominate (brachiocephalic) artery
8. Aorta
9. Ductus arteriosus

10. Superior vena cava (cut)
11. Inferior vena cava
12. Subclavian artery
13. Vertebral artery
14. Transverse scapular artery
15. Axillary artery
16. Radial nerve
17. Thoracodorsal nerve
18. Dorsal rami of thoracic nerves
19. Lung
20. Right and left auricles

21. Heart
22. Continuation of sympathetic trunk
23. Azygous vein
24. Diaphragm
25. Rib with costal artery and vein
26. Ductus venosus
27. Kidney
28. Hilum of left lung (with bronchi and blood vessels cut)

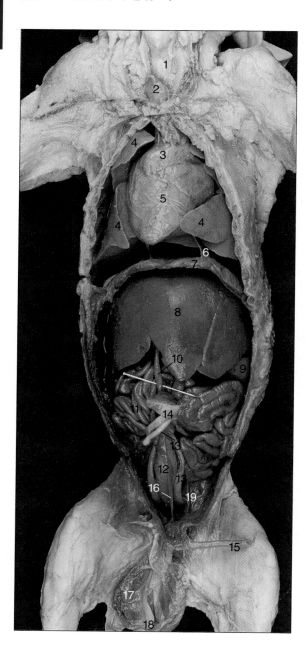

Figure 4-42
Thoracic and Abdominal Viscera of the Fetal Pig

1. Trachea
2. Thyroid gland
3. Thymus
4. Lung
5. Heart in pericardium
6. Mediastinal membrane
7. Diaphragm
8. Liver
9. Spleen
10. Umbilical vein (on pin)
11. Small intestine
12. Urinary bladder
13. Umbilical arteries
14. Skin of umbilicus
15. Penis
16. Urethra
17. Testis
18. Epididymis
19. Spermatic cord (contains spermatic artery and vas deferens, which curves to pass behind base of urinary bladder)

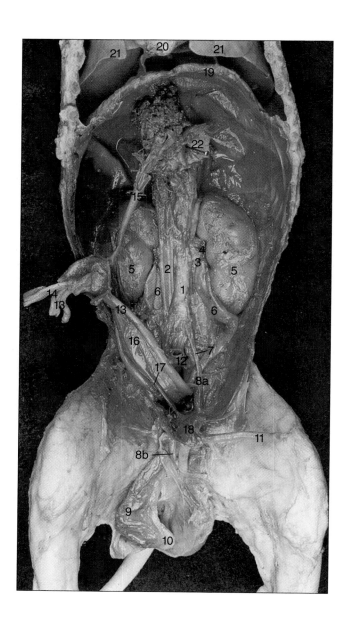

Figure 4-43
Abdominopelvic Cavity of the Fetal Pig, Male, Digestive Viscera Removed

1. Abdominal aorta
2. Abdominal vena cava
3. Renal artery
4. Renal vein
5. Kidney
6. Ureter
7. Spermatic artery
8. a. Vas deferens
   b. Vas deferens (in spermatic cord)
9. Testis
10. Epididymis
11. Penis
12. Rectum (cut)
13. Umbilical arteries
14. Umbilical vein
15. Ductus venosus (only remnants of liver remain)
16. Urinary bladder
17. Urethra
18. Prostate gland
19. Diaphragm
20. Heart
21. Lung
22. Pylorus of stomach (pin in antrum)

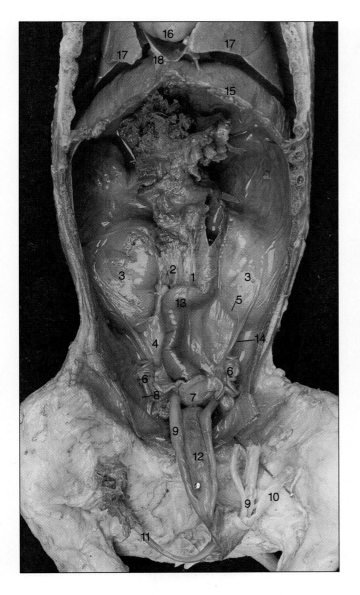

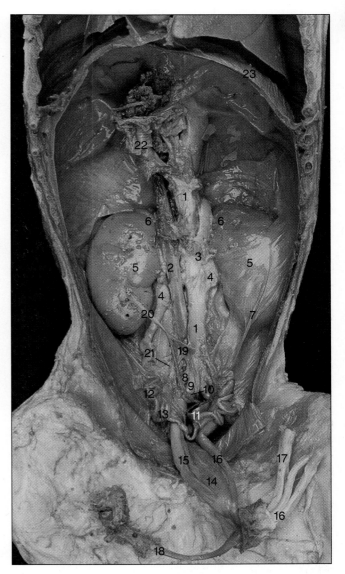

Figure 4-44
Abdominopelvic Cavity of the Fetal Pig, Female, Digestive Viscera Removed

1. Abdominal aorta
2. Abdominal vena cava
3. Kidney (behind intact peritoneum)
4. Ureter (behind intact peritoneum)
5. Ovarian artery
6. Ovary
7. Uterus
8. Uterine horn
9. Umbilical arteries
10. Umbilical vein (lying on pin)
11. Ductus venosus
12. Urinary bladder (reflected and pinned)
13. Sigmoid colon
14. Suspensory ligament of ovary
15. Diaphragm
16. Heart
17. Lung
18. Mediastinal membrane

Figure 4-45
Deep Anatomy of the Abdominopelvic Cavity of the Fetal Pig, Abdominal Viscera Removed, Female

1. Abdominal aorta
2. Abdominal vena cava
3. Renal vein
4. Ureter
5. Kidney (left kidney behind peritoneum)
6. Adrenal gland
7. Suspensory ligament of ovary
8. External iliac artery
9. Internal iliac artery
10. Median sacral (caudal) artery
11. Rectum (cut)
12. Ovary
13. Uterine horn
14. Urinary bladder (reflected)
15. Urethra
16. Umbilical arteries
17. Umbilical vein
18. Ductus venosus
19. Posterior (inferior) mesenteric artery
20. Colic artery (pinned to kidney for clarity due to missing colon)
21. Superior hemorrhoidal artery
22. Remnant of small intestine (duodenum)
23. Diaphragm

Figure 4-46
**General Anatomy of the Male Rat,
Abdominal Cavity Exposed, Ventral View**

1. Thorax
2. Abdomen
3. External oblique muscle (reflected and pinned)
4. Internal oblique muscle (lying on pin)
5. Transversus abdominis
6. Rectus abdominis
7. Peritoneum
8. Inferior epigastric artery
9. Liver
10. Spleen
11. Kidney
12. Rectum
13. Abdominal fat (small intestine not visible in this photograph)
14. Testis
15. Penis
16. Sternum (xiphoid process)

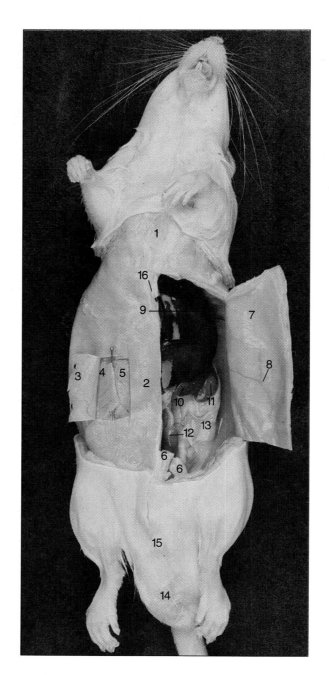

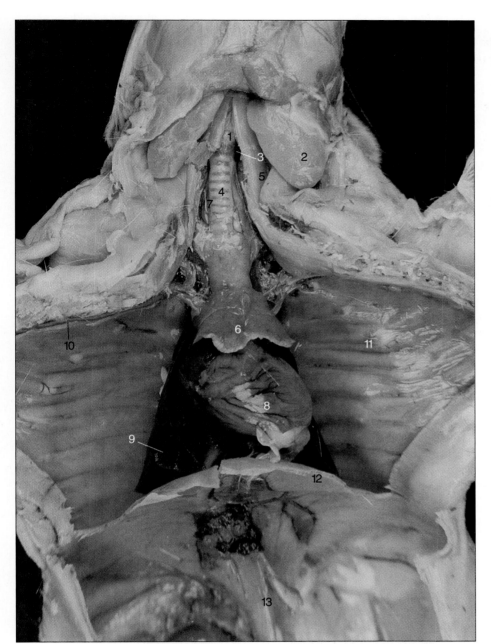

Figure 4-47
Deep Anatomy of the Rat, Neck and Thorax

1. Larynx
2. Salivary gland
3. Thyroid gland
4. Trachea
5. Sternohyoid muscle (unavoidably damaged on animal's right side during vascular perfusion)
6. Thymus
7. Common carotid artery
8. Heart
9. Lung
10. Internal mammary vein
11. Rib and intercostal artery and vein
12. Diaphragm
13. Crus of diaphragm

Figure 4-48
Abdominopelvic Cavity of the Male Rat

1. Sternum (xiphoid process)
2. Stomach
3. Liver
4. Small intestine (duodenum)
5. Pancreas
6. Spleen
7. Kidney
8. Small intestine (jejuneum and ileum)
9. Large intestine (cecum)
10. Rectum
11. Abdominal fat
12. Urinary bladder
13. Rectus abdominis muscle (cut)
14. Testis in scrotum
15. Epididymis
16. Penis
17. Seminal vesicle
18. Cremasteric fascia

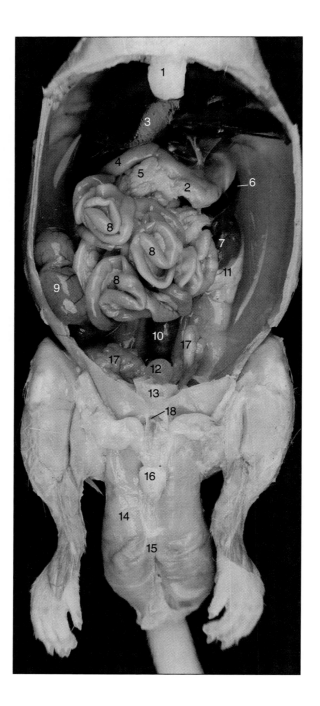

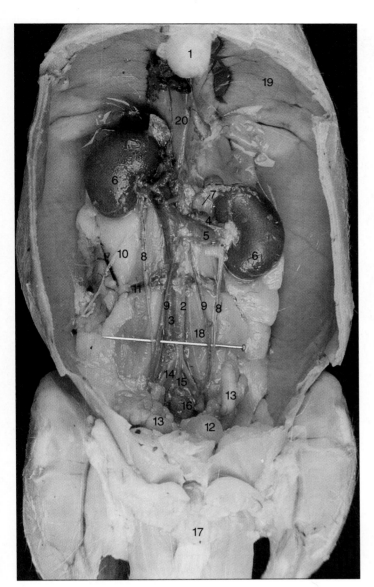

Figure 4-49
Abdominopelvic Cavity of the Rat, Male,
Digestive Viscera Removed
1. Sternum (xiphoid process)
2. Abdominal aorta
3. Abdominal vena cava
4. Renal artery
5. Renal vein
6. Kidney
7. Adrenal gland
8. Ureter (lying on pin)
9. Spermatic artery (lying on pin)
10. Lumbar nerve (medial branch, extended for clarity)
11. Iliolumbar artery and vein
12. Urinary bladder
13. Seminal vesicle
14. Common iliac artery
15. Median sacral (caudal) artery
16. Rectum (cut)
17. Penis
18. Psoas muscle
19. Diaphragm
20. Crus of diaphragm

Figure 4-50
Abdominopelvic Cavity of the Rat, Female, Digestive Viscera Removed

1. Abdominal aorta
2. Abdominal vena cava
3. Renal artery
4. Renal vein
5. Ureter (lying on pin)
6. Iliolumbar artery
7. Iliolumbar vein
8. Rectum (cut)
9. Ovary
10. Uterus
11. Uterine horn
12. Urinary bladder
13. Abdominal fat
14. Ovarian artery and vein
15. Crus of diaphragm
16. Kidney
17. Adrenal gland

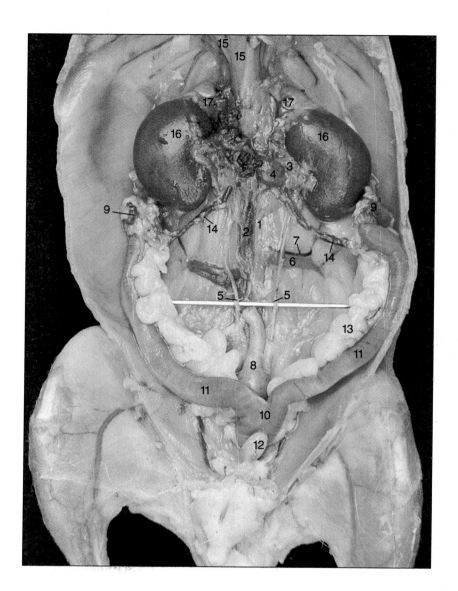

CHAPTER

5

Reference Tables

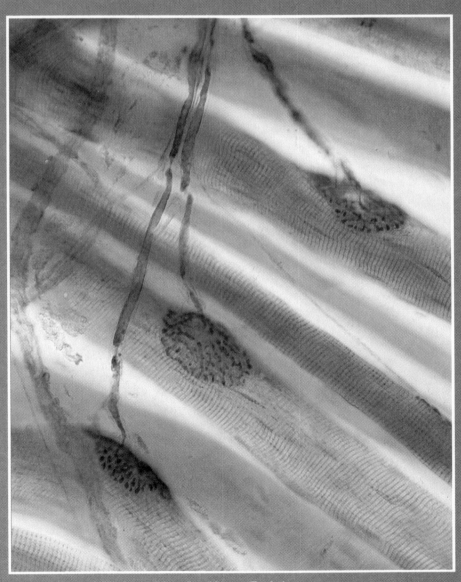

Innervation of Skeletal Muscle: Motor Endplate

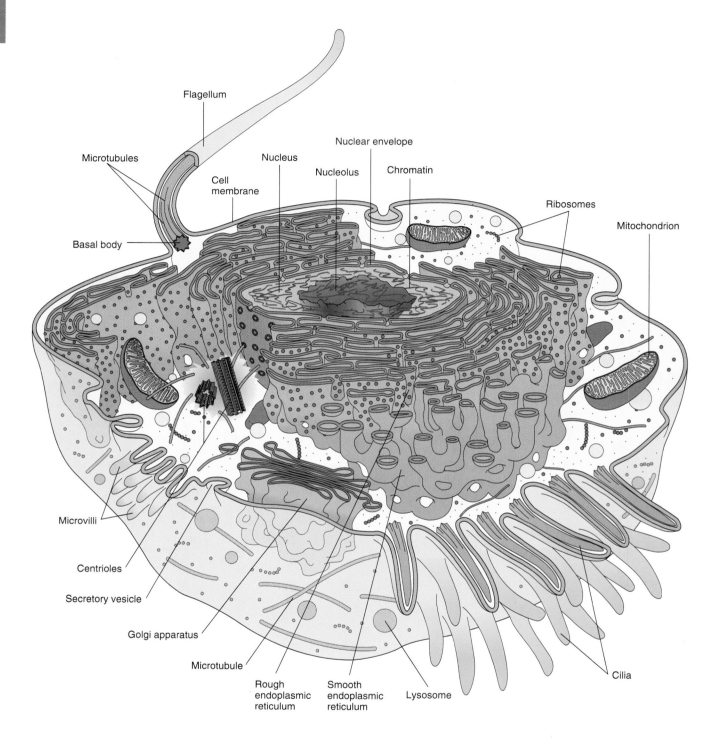

Figure 5-1
Generalized illustration of a cell.

TABLE 5.1 — Structure and Function of Some Cellular Components

STRUCTURE	DESCRIPTION AND FUNCTION
MEMBRANOUS	
Plasma membrane	Composed mainly of phospholipid bilayer with globular proteins floating dynamically on, in, and through it. Separates living cell contents from nonliving environment. Maintains cellular integrity. Embedded molecules serve as identifying cell markers (antigens), receptor molecules for hormones and related substances, signal transducers, selective ion channels, and transporter mechanisms.
Endoplasmic reticulum	Complex of membranous canals, sacs, and vesicles. Transports material within the cell; provides attachment for ribosomes; contributes to synthesis of lipids, steroids, and some carbohydrates used to form glycoproteins.
Golgi apparatus	Flattened membranous sacs. Synthesizes and packages carbohydrates and glycoproteins.
Lysosomes	Small membranous sacs. Contains enzymes used in intracellular digestion.
Peroxisomes	Small membranous vesicles. Contains peroxidase enzymes used in breakdown of complex toxins and other organic molecules.
Mitochondria	Small membranous sacs with complex internal structure and separate DNA. Contains enzymes of Krebs cycle; central to carbohydrate metabolism and synthesis of ATP.
Nucleus	Nuclear contents, notably DNA, separated from cytoplasm by porous nuclear envelope.
NONMEMBRANOUS	
Ribosomes	Small structures composed of two parts containing protein and RNA molecules. Often associated with endoplasmic reticulum. Synthesizes proteins under instructions of messenger RNA triplet code.
Centrosome	Double structure composed of two, short, rod-like centrioles. Important in distribution of chromosomes during cell division and in formation of cilia.
Microfilaments and microtubules	Composed of protein complexes. Acts as cytoskeletal framework. Functions in whole-cell and local membrane movements, cellular elasticity, and formation of cellular extensions (e.g., microvilli).
Cilia and flagella	Movable membranous extensions. Important in movement of fluid environment over stationary cell surface (cilia) and cell itself (flagellum of sperm cell).
Nucleolus	Dense object composed of protein and RNA molecules. Essential in ribosome formation.

TABLE 5.2 — Some Membrane Transport Processes

PROCESS	DESCRIPTION
PHYSICAL PROCESSES: DO NOT REQUIRE LOCAL EXPENDITURE OF METABOLIC ENERGY	
Bulk flow	Movement of substances from higher pressures toward lower pressures. Examples: movement of gases in and out of ventilatory tree during breathing, movement of blood through arteries and veins due to pumping action of heart.
Diffusion	Movement of ions or molecules from higher concentrations toward lower concentrations due to random molecular collisions. Examples: movement of sodium and potassium ions and glucose molecules in extracellular fluid.
Filtration	Bulk flow through a semipermeable membrane. Example: movement of fluid and small molecules through kidney capillary walls due to hydrostatic pressure.
Dialysis	Diffusion of solute molecules through a semipermeable membrane. Example: passage of lipid-soluble substances, such as steroid molecules, through cell membrane.
Osmosis	Diffusion of water down *its* concentration gradient through a semipermeable membrane. Osmosis generally operates *against* concentration gradient of solute(s) to which the membrane is *im*permeable. Example: net movement of extracellular fluid into the venous ends of capillaries under influence of *im*permeant plasma proteins.
Facilitated diffusion	Diffusion through an otherwise impermeable membrane by means of carrier molecules. Example: movement of glucose through muscle cell membranes (requires insulin to enhance action of facilitating carriers).
PHYSIOLOGICAL PROCESSES: REQUIRE LOCAL EXPENDITURE OF METABOLIC ENERGY	
Active transport	Carrier-mediated transport of ions or molecules through a living membrane via energy-requiring shape change of carrier molecule. Energy expenditure permits transport from lower to higher concentration. Examples: movement of sodium from inside to outside of resting nerve cells; transport of potassium and calcium from outside to inside cells, thereby causing high internal concentrations of these ions.
Phagocytosis and pinocytosis	Transport of large particles or fluid into a cell via engulfing action of membrane followed by pinching off to form an intracellular vesicle. Both are processes of endocytosis. Example: trapping of bacteria by white blood cells.
Exocytosis	Transport of substances out of a cell by fusion of internal vesicle with cell membrane and release of contents to the exterior. Examples: secretion of hormones and neurotransmitters, such as prolactin and acetylcholine.

TABLE 5.3   Terms for Bone Structure

TERM	DEFINITION
Epiphysis	Either rounded end of head of a long bone
Diaphysis	The shaft of a long bone
Anatomic neck	The epiphyseal growth plate
Surgical neck	The narrow part of a long bone, just past the head, where fracture is most likely
Ramus	A branch
Cornu	A horn
Hamulus	A hook
Lingula	A tongue
Foramen (pl. foramina)	A hole; an opening into or through a bone to permit passage of blood vessels, nerves, or ligaments
Fossa	A valley; a relatively deep pit or depression
Fovea	A relatively small pit or depression
Sulcus	A narrow valley
Meatus	A tunnel
Trochanter	A large, blunt, rounded process that serves as a site for muscle attachment
Tubercle	A small, blunt, rounded process that serves as a site for muscle attachment
Tuberosity	A large, rounded, often rough eminence or surface that serves as a site for muscle attachment
Condyle	A large, rounded process at the end of a bone, usually contributing to a joint
Epicondyle	A smaller, rounded process at the end of a bone, on top of a condyle, usually contributing to a joint
Trochlea	A pulley; a smooth notched surface often found at a joint
Facet	A face; a smooth, nearly flat surface at a joint
Fissure	A crack or cleft
Crest or crista	A narrow ridge
Spine	A pointed ridge
Fontanel	Specifically, six spaces between the cranial bones of the fetal and infant skull prior to closure of the sutures
Second and fifth intercostal spaces	Specifically refers to a place between the 2nd and 3rd rib and a place between the 5th and 6th ribs where the second and first heart sounds, respectively, can be heard especially well

TABLE 5.4 — Bones of the Human Skeleton

PART OF THE BODY	NAMES OF BONES
AXIAL SKELETON (80 BONES TOTAL)	
Skull (28 bones)	
Cranium (8 bones)*	Frontal (1) Parietal (1 pair) Temporal (1 pair) Occipital (1) Sphenoid (1) Ethmoid (1)
Face (14 bones)	Lacrimal (1 pair) Nasal (1 pair) Palatine (1 pair) Inferior nasal conchae (turbinates) (1 pair) Vomer (1) Maxillae (1 pair) Zygomatic (malar) (1 pair) Mandible (1)
Middle ear (6 bones)	Malleus (1 pair) Incus (1 pair) Stapes (1 pair)
Hyoid bone (1)	
Spinal column (26 bones total)	Cervical vertebrae (7) Thoracic vertebrae (12) Lumbar vertebrae (5) Sacrum (4–5 fused into 1) Coccyx (4–5 fused into 1)
Sternum and ribs (25 bones total)	Sternum (1) True ribs (7 pairs) False ribs (5 pairs)
APPENDICULAR SKELETON (126 BONES TOTAL)	
Shoulder girdle and arm (64 bones total)	Clavicle (1 pair) Scapula (1 pair) Humerus (1 pair) Ulna (1 pair) Radius (1 pair) Carpals (8 pairs; navicular (scaphoid), lunate, triangular (triquetum), pisiform, greater multangular (trapezium), lesser multangular (trapezoid), capitate, hamate) Metacarpals (5 pairs) Phalanges (14 pairs)
Pelvic girdle and leg (62 bones total)	Os coxae (1 pair: 2 innominate bones each formed by fusion of ilium, ischium, and pubis) Femur (1 pair) Patella* (1 pair) Tibia (1 pair) Fibula (1 pair) Tarsals (7 pairs: Talus, calcaneus, navicular, medial cuneiform, intermediate cuneiform, lateral cuneiform, cuboid.) Metatarsals (5 pairs) Phalanges (14 pairs)

*A variable number of rounded bones known as **sesamoid bones** (because of their supposed resemblance to sesame seeds) may appear in various tendons, especially those in the wrist, knee, ankle, and foot. Only two of them, the patellae, are commonly found. **Wormian bones** are found in variable numbers within the suture lines of the skull. While most are commonly smaller than the size of fingernails, some can be surprisingly large.

TABLE 5.5 — Comparison of Female and Male Skeletons

Differences between male and female skeletons are graded, not discrete. Female skeletons can have many masculine features, and vice versa. Nevertheless, there are trends, including those listed below. A typically masculine pelvis is called *android;* a typically feminine pelvis is called *gynecoid.* Many intermediate types exist.

PORTION OF SKELETON	FEMALE	MALE
GENERAL FORM	Bones lighter and thinner / Muscle attachment sites smaller and smoother / Joint surfaces relatively small	Bones heavier and thicker / Muscle attachment sites larger and rougher / Joint surfaces relatively large
PELVIS		
Pelvic cavity	Wider in all dimensions / Shorter and roomier / Pelvic outlet relatively large	Smaller in all dimensions / Deeper / Pelvic outlet usually obstructed
Sacrum	Short, wide, flat concavity more pronounced in a posterior direction; sacral promontory less pronounced	Long, narrow, with smooth concavity of sacral curvature; sacral promontory more pronounced
Coccyx	More movable and follows posterior direction of sacral curvature	Less movable
Pubic arch	Greater than a 90° angle	Less than a 90° angle
Ischial spine, ischial tuberosity, and anterior super iliac spine	Oriented outward and further apart	Oriented inward
Greater sciatic notch	Narrow	Wide

TABLE 5.6 — Extrinsic Muscles of the Eye

NAME	ORIGIN	INSERTION	ACTION	INNERVATION
Rectus superior	Tendinous ring of tissue which surrounds optic foramen at back of orbit	Top of eyeball	Rolls eye upward	Oculomotor
Rectus inferior		Bottom of eyeball	Rolls eye downward	Oculomotor
Rectus lateralis		Lateral side of eyeball	Rolls eye laterally	Abducens
Rectus medius		Medial side of eyeball	Rolls eye medially	Oculomotor
Obliquus superior		Top of eyeball under rectus superior, through trochlea	Prevents rotation of eyeball on axis; directs gaze down and laterally	Trochlear
Obliquus inferior	Maxilla at front of orbit	Lateral side of eyeball under rectus lateralis	Prevents rotation of eye on axis; directs gaze up and laterally	Oculomotor

TABLE 5.7   Facial Muscles

NAME	ORIGIN	INSERTION	ACTION	INNERVATION
Buccinator	Maxillary and mandibular alveolar processes	Into orbicularis oris at sides of mouth	Compresses cheek, retracts corner of mouth as in playing a brass musical instrument	Facial
Orbicularis oris	Maxillae, mandible, nasal septum	Fibers encircle mouth, insert on fascia	Puckering, shaping of mouth in speech	Facial
Orbicularis oculi	Maxillae, frontal bone	Fibers encircle orbit	Closes eye, assists in squinting	Facial
Epicranius (Occipitofrontalis)	Occipital bone	Skin around eyebrows and above nose	Moves scalp, elevates eyebrows	Facial
Zygomaticus major	Zygomatic bone	Into orbicularis oris at corners of mouth	Retracts and elevates corners of mouth as in smiling	Facial
Zygomaticus minor	Zygomatic bone	Into orbicularis oris of upper lip	Elevates upper lip, assists in smiling	Facial
Levator palpebrae superioris	Lesser wing of sphenoid	Skin of upper eyelid	Elevates upper eyelid	Oculomotor
Corrugator supercilii	Bridge of nose, orbicularis oculi	Skin of eyebrows	Depresses and adducts eyebrows; furrows forehead as in frowning	Facial

TABLE 5.8   Chewing Muscles

NAME	ORIGIN	INSERTION	ACTION	INNERVATION
Masseter	Zygomatic arch and maxilla	Lateral surface of mandible	Closes jaw	Trigeminal
Temporalis	Temporal bone	Coronoid process of mandible	Closes jaw	Trigeminal
Pterygoid (medial and lateral)	Pterygoid processes of sphenoid bone	Medial surface of mandible	Moves jaw from side to side; grates teeth for chewing	Trigeminal

TABLE 5.9 — Muscles of the Throat

NAME	ORIGIN	INSERTION	ACTION	INNERVATION
Digastric	Mastoid process of temporal bone	Mandible	Elevates hyoid bone; depresses and retracts mandible	Posterior portion: Facial Anterior portion: Mandibular branch of trigeminal
Mylohyoid	Mandible	Hyoid	Elevates floor of mouth when mandible is fixed; depresses mandible when hyoid is fixed	Mandibular division of trigeminal
Omohyoid	Superior border of scapula and tendon from clavicle	Hyoid	Depresses hyoid; stabilizes hyoid when opening mouth	C1–C3 via ansa hypoglossi
Sternohyoid	Manubrium of sternum; costal cartilage 1	Hyoid	Depresses hyoid; stabilizes hyoid when opening mouth	C1–C3 via ansa hypoglossi
Sternothyroid	Manubrium of sternum; costal cartilages 1 and 2	Thyroid cartilage	Depresses larynx; stabilizes larynx when opening mouth	Upper cervical nerves via ansa cervicalis and ansa hypoglossi

TABLE 5.10 — Muscles of the Tongue

NAME	ORIGIN	INSERTION	ACTION	INNERVATION
Intrinsic muscles: Longitudinal, vertical, and transverse	Within tongue	Within tongue	Change shape of tongue in speaking, chewing, licking	Hypoglossal
Genioglossus	Genu of mandible	Tongue	Depresses and protrudes tongue	Hypoglossal
Hyoglossus	Hyoid	Side of tongue	Depresses and retracts tongue	Hypoglossal
Styloglossus	Styloid process of temporal bone	Inferior and lateral aspect of tongue	Retracts tongue	Hypoglossal

NOTE: The three above-named muscles are **extrinsic muscles of the tongue**, so identified because their origins lie outside the muscular tongue itself.

TABLE 5.11   Muscles of the Pharynx and Palate

NAME	ORIGIN	INSERTION	ACTION	INNERVATION
Constrictor pharyngis inferior	Cricoid and thyroid cartilages	Median raphe of pharynx	Constricts lower pharynx during swallowing	Glossopharyngeal and vagus
Constrictor pharyngis medius	Greater and lesser cornu of hyoid	Median raphe of pharynx	Constricts middle pharynx during swallowing	Glossopharyngeal and vagus
Constrictor pharyngis superior	Middle pterygoid plate, mandible, floor of mouth	Median raphe of pharynx	Constricts upper pharynx during swallowing	Glossopharyngeal and vagus
Stylopharyngeus	Styloid process of temporal bone	Sides of pharynx; thyroid cartilage	Elevates and dilates pharynx	Glossopharyngeal
Palatopharyngeus	Soft palate	Pharynx	Narrows fauces; depresses palate; elevates pharynx	Glossopharyngeal and vagus
Palatoglossus	Soft palate	Tongue	Narrows fauces; elevates back of tongue	Pharyngeal plexus
Levator veli palatini	Temporal bone and cartilage of Eustachian tube	Soft palate	Elevates soft palate	Glossopharyngeal and vagus
Tensor veli palatini	Sphenoid bone and cartilage of Eustachian tube	Soft palate	Increases tension of soft palate; opens Eustachian tube as in yawning	Mandibular division of trigeminal

NOTE: The palatopharyngeus muscle and its mucous membrane covering form the clearly seen arch of the soft palate, from which hangs the uvula. Just anterior to this arch on each side is the palatoglossus muscle which, with its mucous membrane covering, forms the more lateral and less clearly seen glossopalatine arch. Between these two arches on each side is a fossa that houses the lymph node known as the palatine tonsil.

TABLE 5.12   Muscles That Move the Head

NAME	ORIGIN	INSERTION	ACTION	INNERVATION
Sternocleidomastoid	Sternum and clavicle	Mastoid process of temporal bone	Bows head, rotates head	Spinal accessory, C2–C3
Trapezius	Acromial process of clavicle and spine of scapula	Occipital bone, ligamentum nuchae, spines of 7th cervical and all thoracic vertebrae	Extends head, rotates head	Spinal accessory, C3–C4
Obliquus capitis inferior	Spinous process of axis	Transverse process of atlas	Rotates head	Branch of suboccipital
Splenius capitis	Ligamentum nuchae, spines of 7th cervical and top four thoracic vertebrae	Occipital bone and mastoid process of temporal bone	Extends head, rotates head	Middle and lower cervical spinal nerves
Semispinalis capitis	See MUSCLES OF THE VERTEBRAL COLUMN. The capitis division of this muscle inserts on the occipital bone. When the vertebrae serve as the origin and the occipital bone as the insertion, this muscle (bilaterally) extends the head or (unilaterally) draws the head toward the contracting side.			
Longissimus capitis	See MUSCLES OF THE VERTEBRAL COLUMN. The capitis division of this muscle inserts on the mastoid process of the temporal bone. When the vertebrae serve as the origin and the occipital bone as the insertion, this muscle (bilaterally) extends the head or (unilaterally) draws and rotates the head toward the contracting side.			

TABLE 5.13 — Muscles That Move the Shoulder

NAME	ORIGIN	INSERTION	ACTION	INNERVATION
Trapezius	See MUSCLES THAT MOVE THE HEAD. If origin and insertion are reversed, this muscle causes elevation of shoulders, as in shrugging, by elevating clavicle and scapula.			
Pectoralis minor	Outer surface of third, fourth, and fifth ribs	Coracoid process of scapula	Depresses shoulder, rotates scapula forward and down; can assist in elevating ribs	Long thoracic
Serratus anterior	Outer surface of upper eight or nine ribs	Ventral surface of vertebral border of scapula	Rotates scapula forward and toward thoracic wall; can assist in elevating ribs	Spinal accessory, C3–C4
Rhomboideus major	Spines of second to fifth thoracic vertebrae	Vertebral border of scapula	Adducts scapula, rotates slightly upward	Dorsal scapular
Rhomboideus minor	Spines of seventh cervical and first thoracic vertebrae	Vertebral border of scapula	Adducts scapula	Dorsal scapular

NOTE: The **triangle of auscultation** is formed at the caudal medial border of the scapula by the edges of the latissimus dorsi, trapezius, and rhomboideus muscles.

TABLE 5.14 — Muscles That Move the Upper Arm

NAME	ORIGIN	INSERTION	ACTION	INNERVATION
Pectoralis major	Clavicle, sternum, cartilages of second to sixth ribs	Crest and greater tubercle of humerus	Flexes and adducts arm	Anterior thoracic
Latissimus dorsi	Spinous processes of lower six thoracic and all lumbar vertebrae, sacral spine, iliac crest and lower four ribs	Intertubercular groove of humerus	Extends, adducts, rotates arm medially, draws shoulder down and back	Thoracodorsal
Deltoideus	Clavicle and acromion and spine of scapula	Deltoid tuberosity of humerus	Abducts arm	Axillary
Coracobrachialis	Coracoid process of scapula	Medial surface of humerus	Adducts arm; assists in flexion and medial rotation	Musculocutaneous
Teres major	Medial border of scapula	Just distal to lesser tubercle of humerus	Adducts, extends, rotates arm medially	Lower subscapular
Teres minor	Medial border of scapula	Greater tubercle of humerus	Rotates arm laterally	Axillary
Subscapularis	Subscapular fossa of scapula	Lesser tubercle of humerus	Extends and medially rotates arm	Subscapular C5, C6
Supraspinatus	Supraspinous fossa of scapula	Greater tubercle of humerus	Initiates abduction of arm	Suprascapular C5, C6
Infraspinatus	Infraspinous fossa of scapula	Greater tubercle of humerus	Extends and laterally rotates arm	Suprascapular C5, C6

NOTE: The **rotator cuff** is formed from the tendons of the last four muscles named above because together they form a cuff that binds the humerus into the shallow glenoid fossa. A rotator cuff injury involves damage to one or more of these muscles or their tendons.

NOTE: Alone, the deltoid cannot initiate the first 15° of abduction, which is a duty of the supraspinatus muscle and its innervation and which is separate from that of the deltoid. Differential assessment of peripheral nerve injury is possible by asking a patient to abduct the arm from anatomical position.

TABLE 5.15   Muscles That Move the Lower Arm

NAME	ORIGIN	INSERTION	ACTION	INNERVATION
Biceps brachii	Long head: Tuberosity above glenoid cavity of scapula Short head: Coracoid process of scapula	Radial tuberosity	Flexes and supinates arm and forearm	Musculocutaneous
Brachialis	Anterior surface of distal humerus	Tuberosity and coronoid process of ulna	Flexes forearm	Musculocutaneous, radial, and median
Brachioradialis	Supracondyloid ridge of humerus	Proximal to styloid process of radius	Flexes forearm	Radial
Triceps brachii	Long head: Infraglenoid tuberosity of scapula Lateral head: Posterior surface of humerus above radial groove Medial head: Posterior surface of humerus below radial groove	Olecranon process of ulna	Extends forearm	Radial
Anconeus	Lateral epicondyle of humerus	Olecranon process and proximal one-fourth of ulna	Extends forearm	Radial
Pronator teres	Medial epicondyle of humerus, coronoid process of ulna	Middle third of lateral surface of radius	Pronates and flexes forearm	Median
Pronator quadratus	Distal shaft of ulna	Distal shaft of radius	Pronates forearm	Median
Supinator	Lateral epicondyle of humerus, proximal end of ulna	Proximal third of radius	Supinates forearm	Median

TABLE 5.16   Muscles That Move the Wrist and Hand

NAME	ORIGIN	INSERTION	ACTION	INNERVATION
FLEXORS				
Flexor carpi ulnaris	Ulna; medial epicondyle of humerus	Fifth metacarpal; pisiform and hamate	Flexes and adducts wrist; flexes forearm	Ulnar
Palmaris longus	Medial epicondyle of humerus	Palmar fascia	Tenses palmar fascia; flexes wrist	Median
Flexor carpi radialis	Medial epicondyle of humerus	First and second metacarpals	Flexes and abducts wrist	Median
Flexor digitorum profundus	Ulna	Distal phalanges 2–5	Flexes fingers and wrist	Median and ulnar
Flexor digitorum superficialis	Medial epicondyle of radius	Middle phalanges 2–5	Flexes fingers and wrist	Median
Flexor pollicis longus	Radius	Distal phalanx of thumb	Flexes thumb and wrist	Median
EXTENSORS				
Extensor carpi ulnaris	Ulna; lateral epicondyle of humerus	Metacarpal 5	Extends hand; adducts little finger	Radial
Extensor digitorum	Lateral epicondyle of humerus	Phalanges 2–5	Extends fingers and wrist	Radial
Extensor carpi radialis brevis	Lateral epicondyle of humerus	Metacarpal 3	Extends and abducts wrist	Radial
Extensor carpi radialis longus	Lateral supracondylar ridge of humerus	Metacarpal 2	Extends and abducts wrist	Radial
Extensor indicis	Ulna	Phalanx 2	Extends forefinger and wrist	Radial
Abductor pollicis longus	Posterior ulna and radius; interosseous membrane	Metacarpal 1	Abducts and extends thumb and wrist	Radial
Extensor pollicis longus	Dorsal surface of ulna	Base of thumb, second phalanx	Extends end of thumb	Radial
Extensor pollicis brevis	Dorsal surface of radius	Dorsal surface of thumb, first phalanx	Extends and abducts thumb; abducts wrist	Posterior interosseous

NOTE: These last two muscles cross the lateral surface of the wrist to form the **anatomical snuff box**. Extend the thumb laterally to see this structure. The radial artery passes through the snuff box; the pulse can be felt there.

TABLE 5.17 — Muscles That Move the Chest Wall: Breathing

NAME	ORIGIN	INSERTION	ACTION	INNERVATION
NOTE: These muscles are overlaid by the latissimus dorsi, trapezius, and the pectoralis, which are functionally part of the appendicular muscle division.				
External intercostals	Inferior border of rib	Superior border of rib	Draws adjacent ribs together	Intercostal
Internal intercostals	Inferior border of rib	Superior border of rib	Draws adjacent ribs together	Intercostal
Transversus thoracis	Lower one third of sternum	Costal cartilage of true ribs (except first rib)	Depresses ribs in exhalation	Intercostal
Diaphragm	Xiphoid process, costal cartilages of lowest six ribs, lumbar vertebrae	Central tendon	Depresses floor of thoracic cavity in inhalation	Phrenic
Sternocleidomastoid	See MUSCLES THAT MOVE THE HEAD. If head acts as origin, then this muscle acts to elevate sternum and rib cage.			
Scalenes	Transverse processes of second to seventh cervical vertebrae	First two ribs	Elevates ribs in inhalation	C5–C8
Levatores costarum	Transverse processes of seventh cervical and first eleven thoracic vertebrae	Angle of rib immediately below origin	Elevates ribs in inhalation	Intercostal

TABLE 5.18 — Muscles That Move the Abdominal Wall

NAME	ORIGIN	INSERTION	ACTION	INNERVATION
External oblique	Lower eight ribs	Iliac crest, linea alba	Compresses abdominal contents	Intercostals 8–12, iliohypogastric, ilioinguinal
Internal oblique	Iliac crest, inguinal ligament, lumbodorsal fascia	Costal cartilages of last three or four ribs	Compresses abdominal contents	Same as external oblique
Transversus abdominis	Iliac crest, inguinal ligament, lumbar fascia, costal cartilages of last six ribs	Xiphoid process, linea alba, pubis	Compresses abdominal contents	Intercostals 7–12, iliohypogastric, ilioinguinal
Rectus abdominis	Pubic crest, symphysis pubis	Xiphoid process, costal cartilages of fifth, six, and seventh ribs	Flexes trunk, compresses abdominal contents	Intercostals 7–12

TABLE 5.19   Muscles of the Pelvic Floor: The Pelvic Diaphragm

NAME	ORIGIN	INSERTION	ACTION	INNERVATION
Levator ani	Posterior surface of pubis, ischial spine	Coccyx	Support pelvic organs. Supports pregnant uterus, participates in childbirth	Pudendal
Coccygeus (posterior continuation of levator ani)	Ischial spine	Coccyx, sacrum	Same as levator ani	Pudendal
Spincter ani externus	Coccyx	Central tendon of perineum	Closes anal canal	Pudendal and S4
Spincter urethrae	Pubic ramus	Central tendon of perineum	Constricts urethra	Pudendal
Ischiocavernosus	Ischial ramus	Corpus cavernosum	Compresses base of penis or clitoris	Pudendal
Transverse perinei	Ischial ramus	Central tendon of perineum	Supports pelvic floor	Pudendal
Bulbospongiosus (male)	Perineum and bulb of penis	Central tendon of perineum	Constricts urethra and erects penis	Pudendal
Bulbospongiosus (female)	Central tendon of perineum	Base of clitoris	Erects clitoris	Pudendal

TABLE 5.20 — Muscles of the Vertebral Column: Muscles of Erect Posture

NAME	ORIGIN	INSERTION	ACTION	INNERVATION
NOTE: Muscles of the abdominal wall function as postural muscles.				
Iliopsoas	Postural muscle when femur acts as origin	See MUSCLES LOCATED IN THE ANTERIOR HIP		

ERECTOR SPINAE GROUP

Composed of three muscle groups, each of which has subgroups. The three major groups are the laterally placed **iliocostalis**, the intermediately placed **longissimus**, and the medially placed **spinalis**.

NAME	ORIGIN	INSERTION	ACTION	INNERVATION
Iliocostalis Lumborum Thoracis Cervicis	Iliac crest and all ribs	Ribs or transverse processes roughly six vertebrae above origin	Extends trunk and neck, maintains erect posture, rotates trunk and neck	Dorsal rami of lumbar, thoracic, and cervical spinal nerves
Longissimus Thoracis Cervicis Capitis	Transverse processes of thoracic and lumbar vertebrae	Transverse processes roughly twelve vertebrae above origin, some ribs, and mastoid process of temporal bone	Extends trunk and neck, maintains erect posture, rotates trunk and head	Dorsal rami of lumbar, thoracic, and cervical spinal nerves
Spinalis Thoracis Cervicis Capitis	Spinous processes of upper lumbar and lower thoracic vertebrae	Spinous processes of upper thoracic vertebrae, cervical vertebrae and occipital bone	Extends trunk	Dorsal rami of lumbar and thoracic spinal nerves
Semispinalis	Transverse processes of seventh cervical and thoracic vertebrae	Spinous processes roughly six vertebrae above origin, occipital bone	Extends and rotates vertebral column and head	Dorsal rami of spinal nerves
Multifidus	Pelvic girdle, lumbar vertebrae, transverse processes of thoracic and lower cervical vertebrae	Spinous processes three vertebrae above origin	Extends and rotates trunk	Dorsal rami of lumbar, thoracic, and cervical spinal nerves
Quadratus lumborum	Posterior iliac crest and lower three lumbar vertebrae	Twelfth rib and transverse processes of top four lumbar vertebrae	Lateral flexion of trunk, pelvic extension	T12, L1

TABLE 5.21 — Muscles Located in the Lateral Hip

NAME	ORIGIN	INSERTION	ACTION	INNERVATION
Tensor fasciae latae	Anterior iliac crest	Through iliotibial band to lateral tibia	Tenses and abducts thigh	Superior gluteal

TABLE 5.22   Muscles Located in the Anterior Hip

NAME	ORIGIN	INSERTION	ACTION	INNERVATION
Iliopsoas Two components: Iliacus and psoas	Transverse processes of lumbar vertebrae, iliac fossa	Lesser trochanter of femur and iliopubic junction	Flexes and laterally rotates thigh, also flexes trunk	L1–L3
Rectus femoris	See MUSCLES LOCATED IN ANTERIOR THIGH			

TABLE 5.23   Muscles Located in the Posterior Hip

NAME	ORIGIN	INSERTION	ACTION	INNERVATION
Gluteus maximus	Posterior iliac crest, sacrum, coccyx	Iliotibial tract and gluteal tuberosity of femur	Extends and rotates thigh laterally	Inferior gluteal
Gluteus medius	Lateral surface of ilium	Greater trochanter of femur	Abducts and rotates thigh medially	Superior gluteal
Gluteus minimus	Lateral surface of ilium	Greater trochanter of femur	Abducts and rotates thigh medially	Superior gluteal
Piriformis	Sacrum	Greater trochanter of femur	Abducts and rotates thigh laterally	S1–S2

TABLE 5.24   Muscles Located in the Anterior Thigh

NAME	ORIGIN	INSERTION	ACTION	INNERVATION
QUADRICEPS FEMORIS GROUP				
Rectus femoris	Anterior inferior iliac spine	Tibial tuberosity via patellar tendon	Flexes thigh and extends leg	Femoral
Vastus lateralis	Greater trochanter and linea aspera	Same as rectus femoris	Extends leg	Femoral
Vastus medialis	Linea aspera of femur	Same as rectus femoris	Extends leg	Femoral
Vastus intermedius (located immediately posterior to rectus femoris)	Anterior surface of femur	Same as rectus femoris	Extends leg	Femoral

TABLE 5.25 — Muscles Located in the Medial Thigh

NAME	ORIGIN	INSERTION	ACTION	INNERVATION
ADDUCTOR GROUP				
Adductor brevis	Inferior pubic ramus	Linea aspera of femur	Adducts, rotates, and flexes thigh	Obturator
Adductor longus	Pubic crest and symphysis pubis	Linea aspera of femur	Adducts, rotates, and flexes thigh	Obturator
Adductor magnus	Ischial tuberosity, ischiopubic ramus	Linea aspera of femur	Adducts, rotates, and flexes thigh	Obturator
Gracilis	Symphysis pubis and pubic arch	Medial surface of tibia	Flexes leg and adducts thigh	Obturator
Pectineus	Pubic spine and iliopubic junction	Pectineal line of femur (distal to lesser trochanter)	Flexes and adducts thigh, rotates thigh laterally	Femoral

TABLE 5.26 — Muscles Located in the Posterior Thigh

NAME	ORIGIN	INSERTION	ACTION	INNERVATION
HAMSTRING GROUP				
Biceps femoris	Long head: Ischial tuberosity; Short head: Linea aspera of femur	Lateral portion of head of fibula, lateral tibial condyle	Flexes leg and extends thigh	Tibial and peroneal
Semitendinosus	Ischial tuberosity	Proximal medial tibia	Flexes leg and extends thigh	Tibial
Semimembranosus	Ischial tuberosity	Medial condyle of tibia	Flexes leg and extends thigh	Tibial

TABLE 5.27	Muscles Located in the Lower Leg			
NAME	**ORIGIN**	**INSERTION**	**ACTION**	**INNERVATION**
Gastrocnemius	Lateral and medial tibial condyles, knee capsule	Calcaneus via Achilles tendon	Plantar flexes foot, flexes leg	Tibial
Soleus	Head of fibula, medial surface of tibia	Calcaneus via Achilles tendon	Plantar flexes foot	Tibial
Plantaris	Linea aspera of femur	Calcaneus via Achilles tendon	Plantar flexes foot, flexes leg	Tibial
Popliteus	Lateral condyle of femur	Posterior tibia	Flexes and medially rotates leg	Tibial
Peroneus brevis	Fibula	Metatarsal 5	Plantar flexes foot	Peroneal
Peroneus longus	Fibula and lateral condyle of tibia	Cuneiform 1; Metatarsal 1	Plantar flexes foot	Peroneal
Flexor hallucis longus	Shaft of fibula	Distal phalanx of great toe	Flexes great toe, plantar flexes foot	Tibial
Flexor digitorum longus	Posterior surface of tibia	Distal phalanges of four lateral toes	Flexes toes, plantar flexes foot	Tibial
Tibialis posterior	Interosseous membrane of tibia and fibula	Several tarsals and metatarsals	Plantar flexes foot	Tibial

NOTE: The tendons of the three preceding flexor muscles pass through the ankle just posterior and inferior to the medial malleolus. From posterior to anterior, the order of these tendons is *T*. posterior, F. *d*igitorum longus, and F. *h*allucis longus, which has led to their being casually referred to as *T*om, *D*ick, and *H*arry.

Extensor hallucis longus	Shaft of fibula, interosseous membrane	Distal phalanx of great toe	Extends great toe, dorsiflexes foot	Deep peroneal
Extensor digitorum longus	Lateral tibial condyle, anterior fibular surface	Middle and distal phalanges of four lateral toes	Extends toes, dorsiflexes foot	Deep peroneal
Tibialis anterior	Lateral condyle and body of tibia	First metatarsal and first cuneiform	Dorsiflexes foot	Deep peroneal
Peroneus tertius	Fibula and interosseous membrane	Metatarsal 5	Dorsiflexes and everts foot	Deep peroneal

TABLE 5.28   The Cranial Nerves

NUMBER AND NAME	EXIT FROM SKULL	FUNCTION
I. Olfactory	Cribriform plate of ethmoid	Sensory: Olfaction. Rhythmic sensitivity follows hormonal cycles in females.
II. Optic	Optic foramen	Sensory: Vision. Probable efferents may regulate retinal metabolism and structural renewal.
III. Oculomotor	Orbital fissure	Motor: Rectus superior, rectus inferior, rectus medius, and obliquus inferior muscles. Sensory: Proprioception. Autonomic (parasympathetic): Muscles of iris, ciliary muscle to control lens.
IV. Trochlear	Orbital fissure	Motor: Obliquus superior muscle. Sensory: Proprioception.
V. Trigeminal		
Ophthalmic branch	Orbital fissure	Sensory: Cornea, upper eyelid, scalp, skin of upper face.
Maxillary branch	Foramen rotundum	Sensory: Palate and upper jaw, teeth and gums, nasopharynx, skin of cheek, lower eyelid, upper lip.
Mandibular branch	Foramen ovale	Sensory: Lower jaw, teeth and gums, anterior two-thirds of tongue, mucous membrane of cheek, skin of lower lip, chin, and ear. Motor: Muscles of chewing, throat, middle ear.
VI. Abducens	Orbital fissure	Motor: Rectus lateralis muscle. Sensory: Proprioception.
VII. Facial	Stylomastoid foramen and internal auditory meatus	Motor: Muscles of facial expression, throat middle ear. Sensory: Proprioception, taste (anterior two-thirds of tongue), palate. Autonomic (parasympathetic): Tear glands, salivary glands, and secretory glands in pharynx.
VIII. Auditory	Internal auditory meatus	Sensory: Hearing (cochlear branch), balance (vestibular branch).
IX. Glossopharyngeal	Jugular foramen	Sensory: Posterior one-third of tongue, posterior pharynx, taste (posterior one-third of tongue), proprioception. Motor: Pharyngeal muscle. Autonomic (parasympathetic): Salivary glands, carotid sinus.
X. Vagus	Jugular foramen	Sensory: Inferior pharynx, larynx, internal organs. Motor: Posterior pharynx, larynx, tongue. Autonomic (parasympathetic): Thoracic and abdominal viscera.
XI. Spinal accessory	Jugular foramen	Motor: Posterior pharynx, sternocleidomastoid, trapezius muscles. Sensory: Proprioception.
XII. Hypoglossal	Hypoglossal canal	Motor: Tongue and throat. Sensory: Proprioception.
XIII. Vomeronasal	Internal to skull	Parts of nasopharynx. May allow desert mammals to sense humidity. Function in humans unknown; may respond to pheromones.

Several mnemonic devices exist to aid remembering the names of the 12 standard cranial nerves in order. The most common (and least bawdy) of these is: "On Old Olympus' Towering Tops, A Finn and German Viewed Some Hops." The recently discovered unpaired thirteenth cranial nerve is not contained in this rhyming couplet.

TABLE 5.29   Spinal Nerves and Their Branches

NERVE		SPINAL COMPONENT	INNERVATION
CERVICAL PLEXUS: C1, C2, C3, C4			
	Superficial cutaneous branches		
	Lesser occipital	C2, C3	Skin of scalp above and behind ear
	Greater auricular	C2, C3	Skin in front of, above, and below ear
	Transverse cervical	C2, C3	Skin of anterior aspect of neck
	Supraclavicular	C3, C4	Skin of upper portion of chest and shoulder
	Deep motor branches		
	Ansa cervicalis		
	Anterior root	C1, C2	Geniohyoid, thyrohyoid, and infrahyoid muscles of neck
	Posterior root	C3, C4	Omohyoid, sternohyoid, and sternothyroid muscles of neck
	Phrenic	C3–C5	Diaphragm
	Segmental branches	C1–C5	Deep muscles of neck (levator scapulae ventralis, trapezius, scalenus, and sternocleidomastoid)
BRACHIAL PLEXUS: C5, C6, C7, C8, T1			
	Axillary	Posterior cord (C5–C6)	Skin of shoulder; shoulder joint, deltoid and teres minor muscles
	Radial	Posterior cord (C5–C8, T1)	Skin of posterior lateral surface of arm, forearm, and hand; posterior muscles of brachium and antebrachium (triceps brachii, supinator, anconeus, brachioradialis, extensor carpi radialis brevis, extensor carpi radialis longus, extensor carpi ulnaris)
	Musculocutaneous	Lateral cord (C5–C7)	Skin of lateral surface of forearm; anterior muscles of brachium (coracobrachialis, biceps brachii, brachialis)
	Ulnar	Medial cord (C8, T1)	Skin of medial third of hand; flexor muscles of anterior forearm (flexor carpi ulnaris, flexor digitorum), medial palm and intrinsic flexor muscles of hand (profundus, third and fourth lumbricales)

T2, T3, T4, T5, T6, T7, T8, T9, T10, T11, T12:

No plexus in these segments; branches run directly to intercostal muscles and skin of thorax.

NERVE		SPINAL COMPONENT	INNERVATION
LUMBOSACRAL PLEXUS: L1, L2, L3, L4, L5, S1, S2, S3, S4, S5			
Lumbar	Iliohypogastric	T12–L1	Skin of lower abdomen and buttock; muscles of anterolateral abdominal wall (external abdominal oblique, internal abdominal oblique, transversus abdominis)
	Ilioinguinal	L1	Skin of upper median thigh, scrotum and root of penis in male and labia majora in female; muscles of anterolateral abdominal wall with iliohypogastric nerve
	Genitofemoral	L1, L2	Skin of middle anterior surface of thigh, scrotum in male and labia majora in female; cremaster muscle in male
	Lateral cutaneous femoral	L2, L3	Skin of anterior, lateral, and posterior aspects of thigh
	Femoral	L2–L4	Skin of anterior and medial aspect of thigh and medial aspect of lower extremity and foot; anterior muscles of thigh (iliacus, psoas major, pectineus, rectus femoris, sartorius) and extensor muscles of leg (rectus femoris, vastus lateralis, vastus medialis, vastus intermedius)
	Obturator	L2–L4	Skin of medial aspect of thigh; adductor muscles of lower extremity (external obturator, pectineus, adductor longus, adductor brevis, adductor magnus, gracilis)
	Saphenous	L2–L4	Skin of medial aspect of lower extremity
Sacral	Superior gluteal	L4, L5, S1	Abductor muscles of thigh (gluteus maximus, gluteus medius, tensor fasciae latae)
	Inferior gluteal	L5–S2	Extensor muscle of hip joint (gluteus maximus)
	Nerve to piriformis	S1, S2	Abductor and rotator of thigh (piriformis)
	Nerve to quadratus femoris	L4, L5, S1	Rotators of thigh (gemellus inferior, quadratus femoris)
	Nerve to internal obturator	L5–S2	Rotators of thigh (gemellus superior, internal obturator)
	Perforating cutaneous	S2, S3	Skin over lower medial surface of buttock
	Posterior cutaneous femoral	S1–S3	Skin over lower lateral surface of buttock, anal region, upper posterior surface of thigh, upper aspect of calf, scrotum in male and labia majora in female
	Sciatic	L4–S3	Composed of two nerves (tibial and common fibular); splits into two portions at popliteal fossa; branches from sciatic in thigh region to "hamstring muscles" (biceps femoris, semitendinosus, semimembranosus) and adductor magnus muscle
	Tibial (sural, medial, and lateral plantar)	L4–S3	Skin of posterior surface of leg and sole of foot; muscle innervation includes gastrocnemius, soleus, flexor digitorum longus, flexor hallucis longus, tibialis posterior, popliteus, and intrinsic muscles of the foot
	Common fibular (superficial and deep fibular)	L4–S2	Skin of anterior surface of the leg and dorsum of foot; muscle innervation includes peroneus tertius, peroneus brevis, peroneus longus, tibialis anterior, extensor hallucis longus, extensor digitorum longus, extensor digitorum brevis
	Pudendal	S2–S4	Skin of penis and scrotum in male and skin of clitoris, labia majora, labia minora, and lower vagina in female; muscles of perineum

TABLE 5.30   Formed Elements of Blood

CELL TYPE	DESCRIPTION (WRIGHT'S STAIN)	NORMAL NUMBER (CELLS/μL OF BLOOD)	FUNCTION
Erythrocytes (Red blood cells, RBC)	7.5μ diameter, biconcave disk, no nucleus	4–6 million	Transport of respiratory gases (O_2 and CO_2)
Leukocytes (White blood cells, WBC)		5,000 to 10,000/mm^3	Aid in defense against infections by microorganisms
Granulocytes			
Neutrophil	12–15μ diameter, multilobed nucleus, small pink-purple granules	3,000–7,000 (65% of total leukocytes)	Phagocytosis; elevated in number during acute infections
Eosinophil	10–14μ diameter, bilobed nucleus, large orange granules	100–400 (3% of total leukocytes)	Destroys antigen-antibody complexes, phagocytosizes parasites, involved in allergic response
Basophil	8–12μ diameter, bilobed large purple granules that may obscure nucleus	20–50 (1% of total leukocytes)	Contains biogenic amines; releases heparin, histamine, other chemicals during inflammatory response
Agranulocytes			
Lymphocyte	5–16μ diameter, round or nucleus, indented, single-lobed nucleus, variable amount of cytoplasm	1,500–3,000 (25% of total leukocytes)	Immune response by direct cellular contact or via antibody production; elevated in infectious mononucleosis; suppressed by steroid therapy
Monocyte	12–20μ diameter, horseshoe-shaped nucleus	100–700 (6% of total leukocytes)	Macrophages; phagocytosis
Platelets	2-4μ, appear as cytoplasmic fragments	25,000 to 500,000	Coagulation

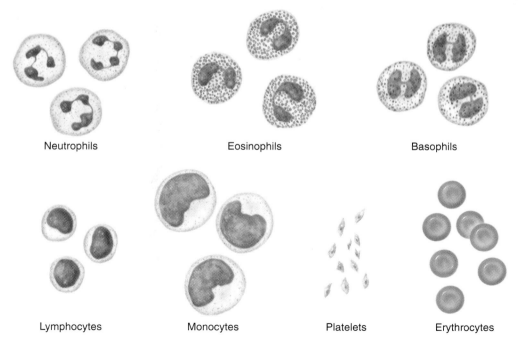

Neutrophils   Eosinophils   Basophils

Lymphocytes   Monocytes   Platelets   Erythrocytes

Figure 5-2
Cellular components in blood.

TABLE 5.31 — Events of the Cardiac Cycle

PHASE	ELECTRICAL EVENTS	MECHANICAL EVENTS	HEART SOUND
Late diastole		AV valves open; semilunar valves closed. Blood enters all chambers by passive filling from venae cavae and pulmonary veins.	
Atrial systole	P wave: SA node depolarizes, wave spreads throughout atria P-R interval: Wave of depolarization reaches SA node. Typical P-R interval (beginning of P-wave to onset of next deviation from baseline) is <0.2 seconds	Atria contracted in response to depolarizing signal. Blood engorges ventricles, adding to stretch of ventricular walls.	
Isometric ventricular contraction	QRS complex: Depolarization of SA node. Bundle of His, and Purkinje fibers spread depolarization through valve ring into ventricular muscle	Ventricles contract in response to depolarizing signal. Papillary muscles relax, allowing AV valves to close. Typically, mitral closure slightly precedes tricuspid closure. Reverberation of blood against valve cusps produces low pitched "lub" of first heart sound. With all valves closed, ventricular pressure rises.	First heart sound (may be split with mitral component preceding tricuspid component).
Ventricular ejection	S-T segment: Entire ventricle is uniformly depolarized	Ejection begins when ventricular pressures exceed back pressures in aorta and pulmonary trunk. Semilunar valves open, blood from this cycle enters aorta and pulmonary trunk. Maintained depolarization during S-T segment permits efficient, coordinated ventricular emptying.	
Isometric ventricular relaxation	T-wave	Repolarization wave spreads through ventricles, permitting relaxation. As ventricular pressures drop below those of aorta and pulmonary trunk, semilunar valves close. Typically, aortic semilunar closes slightly before pulmonary semilunar. Reverberation of blood against closed valve cusps creates higher pitched "dub" of second heart sound. Lowered intraventricular pressures permit papillary muscles to pull AV valves open. Ventricular filling begins.	Second heart sound (typically split with pulmonary component slightly following aortic component, especially during inhalation).

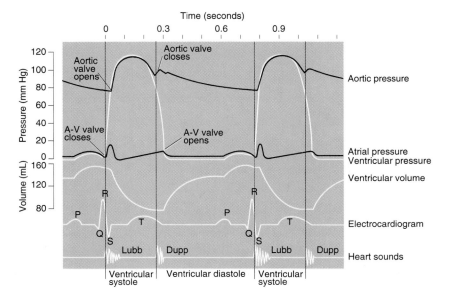

Figure 5-3
A graph of changes that occur in left ventricle during a cardiac cycle.

TABLE 5.32   Major Blood Vessels and Their Branches

MAJOR ARTERY	MAIN BRANCHES
Ascending aorta	Coronary arteries (right and left)
Aortic arch	Innominate (brachiocephalic) Left subclavian Left common carotid
Innominate	Right subclavian Right common carotid
Common carotid (right and left)	Internal carotid External carotid
Subclavian (right and left)	Vertebral (right and left) Axillary (continuation of subclavian)
Axillary	Brachial (continuation of axillary)
Brachial	Radial Ulnar
Radial and ulnar	Palmar arches (superficial and deep)
Circle of Willis	Vertebrals join in cranium to form basilar artery, which then divides to form left and right posterior cerebral arteries. Internal carotids, upon entering cranium, become left and right anterior cerebral arteries. A pair of posterior communicating arteries and an anterior communicating artery join the cerebrals to form an arterial anastomosis, the circle of Willis.
Descending aorta	Intercostal arteries and spinal branches Celiac trunk (branches to hepatic, splenic, and right and left gastric arteries) Mesenteric (superior and inferior) Renal (right and left) Gonadal (spermatic or ovarian; right and left) Parietal branches to diaphragm, dorsal skin and skeletal muscles, and spinal cord Common iliac (right and left)
Common iliac	Internal iliac (or hypogastric; right and left) External iliac (right and left)
External iliac	Femoral (right and left)
Femoral	Popliteal (right and left)
Popliteal	Tibial (anterior and posterior; right and left)
Tibial	Plantar arches

MAJOR VEIN	COMMENT
UPPER EXTREMITY (RIGHT AND LEFT)	
Palmar arch (superficial and deep)	
Medial cubital	Connects cephalic and basilic
Median antebrachial	Median antebrachial and median cubital flow into basilic
Radial and ulnar	Radial and ulnar flow into brachial
Basilic and brachial	Basilic and brachial flow into axillary
Cephalic	Cephalic and axillary flow into subclavian
Axillary (continuation of brachial)	Cephalic and axillary flow into subclavian
Subclavian (continuation of axillary)	Flows into innominate (brachiocephalic)

TABLE 5.32 Major Blood Vessels and Their Branches (continued)

MAJOR VEIN	COMMENT
LOWER EXTREMITY (RIGHT AND LEFT)	
Plantar arch	
Dorsal venous arch	
Anterior tibial	Anterior and posterior tibials unite to form popliteal
Posterior tibial	Anterior and posterior tibials unite to form popliteal
Small saphenous	Flows into popliteal
Popliteal	Popliteal and peroneal unite to form femoral
Peroneal	Popliteal and peroneal unite to form femoral
Femoral	Femoral and great saphenous unite to form external iliac
Great saphenous	Femoral and great saphenous unite to form external iliac
External iliac	External and internal iliacs unite to form common iliac
Internal iliac	External and internal iliacs unite to form common iliac
Common iliac	Flows into inferior vena cava
ABDOMEN	
Lumbar (several pairs)	Flows into inferior vena cava and azygous system
Gonadal (spermatic or ovarian; right and left)	Flows directly into inferior vena cava
Renal (right and left)	Flows directly into inferior vena cava
Suprarenal (adrenal; right and left)	Flows directly into inferior vena cava
Hepatic	Flows directly into inferior vena cava
Mesenteric (superior and inferior)	Flows into hepatic portal system
Splenic	Flows into hepatic portal system
Gastroepiploic (right and left)	Flows into hepatic portal system
Hepatic portal	Conveys blood to liver; hepatic vein flows from liver
THORAX	
Left intercostal	Flows into hemiazygos
Hemiazygos	Flows into azygos
Accessory hemiazygos	Flows into azygos
Right intercostal	Flows into azygos
Azygos	Flows into inferior vena cava
Coronary (right and left)	Flows into right atrium of heart
HEAD AND NECK	
Superior sagittal sinus	
Inferior sagittal sinus	Flows into straight sinus
Straight sinus	Flows into transverse sinus
Cavernous	Flows into petrosal sinus
Petrosal sinus (right and left)	Flows into transverse sinuses
Transverse sinuses (right and left)	Flows into sigmoid sinuses

TABLE 5.32   Major Blood Vessels and Their Branches (continued)

MAJOR VEIN	COMMENT
Sigmoid sinuses (right and left)	Flows into internal jugular vein
Internal jugular	
External jugular	
Vertebral (right and left)	
Innominate (or brachiocephalic; right and left)	Flows into superior vena cava
Superior vena cava	Flows into right atrium of heart

FETAL SYSTEM

Placenta → Umbilical vein → Ductus venosus (bypasses liver) → Inferior vena cava → Right atrium of fetus → Mostly through foramen ovale → Left atrium → Left ventricle → Mostly to fetal head → Return to right atrium → Mostly to right ventricle → Pulmonary trunk → Mostly through ductus arteriosus → Descending aorta → Common iliac arteries → Internal iliac arteries → Umbilical arteries → Placenta

TABLE 5.33   Major Hormones of the Pituitary Gland

HORMONE	CHEMICAL STRUCTURE	TARGET	REGULATION	MAJOR ACTION
ANTERIOR PITUITARY (ADENOHYPOPHYSIS)				
Growth Hormone (GH, Somatotropin)	Protein	General	GH Releasing Hormone from hypothalamus	Enhances protein anabolism, fat catabolism; enhances growth, wound healing, positive nitrogen balance
Prolactin (Prl)	Protein	Breast tissue	Inhibited by dopamine (a prolactin-inhibiting hormone) from hypothalamus	In female, mimics many actions of GH during pregnancy; enhances breast tissue anabolism for lactation
Adrenocorticotropic hormone (ACTH)	Polypeptide	Adrenal cortex	Corticotrophin releasing hormone from hypothalamus	Promotes secretion of glucocorticosteroids by adrenal cortex
Endorphins (several)	Peptide	Central nervous system neurons	Neural activity in hypothalamus in response to stress and probably suckling	Inhibits transmission of pain impulses; enhances feeling of well-being
Thyroid stimulating hormone (TSH)	Glycoprotein	Thyroid gland	Thyroid releasing hormone (TRH) from hypothalamus	Stimulates release of thyroid hormones
Follicle stimulating hormone (FSH)	Glycoprotein	Gonads	Gonadotropin releasing hormone (GnRH) from hypothalamus	Female: Maturation of ovarian follicle; estrogen secretion. Male: Sperm production
Luteinizing hormone (LH)	Glycoprotein	Gonads	Gonadotropin releasing hormone (GnRH) from hypothalamus	Female: Rupture of follicle; ovulation. Male: Testosterone secretion
POSTERIOR PITUITARY (NEUROHYPOPHYSIS)				
Antidiuretic hormone (ADH, Vasopressin)	Peptide	Kidney tubules	Neural activity in hypothalamus in response to brain osmoreceptors; stress	Increase water retention; elevation of blood pressure
Oxytocin	Peptide	Breast tissue, uterus	Neural activity in hypothalamus in response to suckling, uterine stimulation	Let down of milk in lactating breast; uterine smooth muscle contractions

NOTE: The above-named list of hormones is not an exhaustive list of substances now known to be secreted by the pituitary gland. In addition, the listed hormones are known to have several actions, many of which are also not included.

INDEX

A

Abdomen
    in cat, 115–117
    in human, 74–76, 148
    major veins of, 159
    in rat, 130
Abdominal aorta
    in cat, 118–121
    in fetal pig, 128–129
    in rat, 133–134
Abdominal vena cava
    in cat, 118–120
    in fetal pig, 128–129
    in rat, 133–134
Abdominopelvic cavity
    in cat, 120–121
    in fetal pig, 128–129
    in rat, 132–134
Abducens nerve, 141, 154
Abductor digiti minimi, 86
Abductor hallucis, 88
Abductor pollicis longus, 147
Accessory hemiazygos vein, 159
Acetabular notch, 61–62
Acetabulum, 61–62, 93
Achilles tendon, 64
    in cat, 104–106
    in fetal pig, 124
Acromiodeltoid
    in cat, 100–102
    in fetal pig, 122–123, 125
Acromion, 59, 60
Acromiotrapezius, 94, 100–102
ACTH (Adrenocorticotropic
    hormone), 161
Active transport, 138
Adductor brevis, 152
Adductor femoris, 105–106
Adductor longus, 66, 82, 106, 152
Adductor magnus, 67, 82, 84, 152
Adenohypophysis, 23, 161
Adipose tissue, 9
Adrenal cortex, 24
Adrenal gland
    in cat, 118–119, 121
    in fetal pig, 129
    in rat, 133–134
Adrenocorticotropic hormone
    (ACTH), 161
Adrenolumbar artery and
    vein, 121
Adventitia, 28–29, 32, 38
Agranulocyte, 156
Allantoic artery, 119
Alveolar process, 53
Alveoli, 31–32
Anal sphincter, 77
Anaphase, 3
Anatomic neck, defined, 139
Anconeus, 100–101, 146
Android pelvis, versus
    gynecoid, 141
Animal dissection, 90–134
    cat, 90–121

fetal pig, 122–129
    rat, 130–134
Antebrachial fascia, 98, 100
Anterior arch of atlas, 57
Anterior cruciate ligament, 62
Anterior lacrimal crest, 53
Anterior medial malleolar
    artery, 88
Anterior nasal spine, 46–47, 53
Anterior pituitary (adenohypoph-
    ysis), 23, 161
Antidiuretic hormone
    (ADH), 161
Antrum, 42
Anus, 77
Aorta, 8, 114, 127. See also
    Abdominal aorta
Aortic arch, 112–113, 158
Aponeurosis
    epicranial, 68
    external oblique, 74
    of internal oblique, 76
Appendicular skeleton, 140
Appendix, 36
Arcuate artery, 88
Areolar connective tissue, 8
Arm
    bones of, 140
    musculature of, 78–81,
        145–146
    veins of, 158
Artery. See also Specific arteries
    histology, 27–29
    major and main branches
        of, 158
Artherosclerosis, 28
Ascending aorta, 158
Asters, 2–3
Astrocyte, 18
Atlas, 55–57, 91–92
Atrial systole, 157
Auditory nerve, 154
Auricle
    in cat, 112, 114
    in fetal pig, 126–127
Axial skeleton, 140
Axillary artery, 108–109,
    112, 127, 158
Axillary nerve, 96, 99, 114, 145
Axillary vein, 108–109,
    111, 114, 158
Axis, skeletal, 55–57
Axon, 18, 20
Azygos vein, 114, 127, 159

B

Back musculature
    in cat, 101–103
    in human, 71–72
Barr body, 24
Basal body, 136
Basement membrane, 4–7, 29, 39
Basilic vein, 158
Basophil, 25, 156

Basophilic myelocyte, 15
Biceps brachii
    in cat, 109, 112
    in human, 66, 79–80, 99, 146
Biceps femoris
    in cat, 104–106
    in fetal pig, 122, 124
    in human, 67, 83–84,
        86–87, 152
Bifid spinous process, 56–57
Bile duct, 35
Bipolar cell nuclei, 20
Bladder, 7, 39
Blood
    bladder, 39
    components of, 24–27, 156
    histology, 24–27
Blood vessel. See also Artery; Vein
    alveolar, 32
    gallbladder, 36
    large intestine, 34
    listing of major, 158
    lumen, 7
    small intestine, 33
    spleen, 31
Bone. See also Specific bones
    cancellous (spongy), 14
    compact, 14
    descriptive terms for, 139
    developing, 15
    typical structure of long, 64
Bone marrow, 15
Bowman's capsule, 37, 38
Brachial artery, 112, 158
Brachialis, 67, 79–80, 100, 146
Brachial plexus, 109, 114, 155
Brachial vein, 111, 114, 158
Brachiocephalic
    artery, 112–113, 127
Brachiocephalic vein,
    111, 114, 160
Brachioradialis
    in cat, 98–101
    in fetal pig, 123
    in human, 66–67, 80–81, 146
Bregma, 48, 92
Bronchiole, 32
Brush border, 34
Buccinator, 68, 94–95, 142
Bulbospongiosus, 77, 149
Bulbourethral (Cowper's)
    gland, 119
Bulk flow, 138

C

Calcaneal tendon, 67, 86–87, 104
Calcaneus, 64, 87, 90, 93, 104
Calvarium, 48
Camper's fascia, 76
Canaliculus, 14
Cancellous bone, 14, 64
Canine eminence, 53
Canine fossa, 53
Capillary, 29

Capitate carpal, 61
Capitulum, 60
Capsule, parathyroid gland, 22
Cardiac cycle, 157
Cardiac muscle, 17, 65
Carotid artery
    in cat, 97, 111–114
    in fetal pig, 126–127
    in human, 158
    in rat, 131
Carotid canal, 48, 49
Carotid groove, 49
Carpals
    in cat, 90, 92
    in human, 61, 140
Cartilage, 10–11, 15
Cat
    astrocytes, 18
    musculature, 94–109
    skeletal system, 90–93
Caudal artery, 118, 120, 129
Caudal vein, 120
Caudal vertebrae (cat), 90, 92–93
Caudofemoralis, 104, 105
Cavernous vein, 159
Celiac artery, 120–121
Celiac ganglion, 121
Cell
    chief, 33
    component structure and
        function, 137
    generalized illustration
        of, 136
    goblet, 5, 33–34
    granulosa, 42
    hair, 21
    interstitial, 39
    keratinized, 7
    mitotic, 2–3
    parietal, 33
    Purkinje, 18
    Sertoli, 39
    umbrella, 7, 39
Cell body, nerve, 18, 19
Cell membrane, 136
    structure and function
        of, 137
    transport processes, 138
Central canal, 20
Central vein, 35
Centriole, 136
Centrosome, 137
Cephalic vein, 158
    in cat, 98, 100–101, 111
    in fetal pig, 126
Cerebellar fossa, 52
Cerebellum, 18
Cerebral fossa, 52
Cervical nerves, 96, 144, 150
Cervical plexus, 155
Cervical vertebrae
    in cat, 90–92
    in human, 55–56, 140
Chest musculature, 74, 148

Chewing musculature, 142
Chief cell, 33
Chondrocyte, 10–11
Chromatin, 2, 136
Chromosome, 2–3
Cilia, 5–6, 136–137
Ciliated columnar epithelium,
    5–6, 40
Circle of Willis, 158
Circulatory system, 158–160
Clavicle, 59, 69, 79, 140
Clavicular notch, 58, 59
Clavobrachialis
    in cat, 94–95, 98–102,
        107, 115
    in fetal pig, 122–123, 125
Clavotrapezius
    in cat, 94–95, 101–102
    in fetal pig, 122–123, 125
Cleavage furrow, 3
Cleft, pituitary gland, 23
Cleidomastoid, 94–96
Clitoris, 77
Coccygeus, 77, 149
Coccyx, 77, 140–141
Colic artery, 129
Collagen fiber, 8–10
Columnar epithelium, 5–6, 40–41
Common carotid artery, 111–113,
    126–127, 131, 158
Common iliac artery,
    133, 158, 160
Common iliac vein, 120, 159
Compact bone, 14, 45, 64
Condylar canal, 49, 52
Condylar fossa, 49, 52
Condyle, 139
Condyloid process, 53
Cones, 20
Connective tissue, 8–9
Conoid tubercle, 59
Constrictors, pharyngeal, 70, 144
Convoluted tubules, 38
Coracobrachialis
    in cat, 109
    in human, 79, 145
Coracoid process, 59, 60
Core, nerve, 20
Cornified layer, skin, 12
Cornu, 139
Coronal suture
    in cat, 91–92
    in human, 47–48
Corona radiata, 41–42
Coronary vein, 159
Coronoid fossa, 60
Coronoid process, 53, 60
Corpora cavernosa, 41
Corpuscles
    Hassall's, 30
    Meissner's, 12
    Pacinian, 13
    renal, 37
Corpus spongiosum, 39, 41
Corrugator supercilii, 142
Corti, Organ of, 21
Costaclavicular ligament, 59

Costal artery, 127
Costal cartilage, 58, 59
Costal facet, 56
Costal groove, 58
Costal vein, 127
Cranial nerve, 154
Cranium, 48, 90, 140. See also
    Head; Skull
Cremasteric fascia, 132
Crest, 139
    ethmoidal, 52, 53
    frontal, 49
    intertrochanteric, 62
    lacrimal, 53
    nasal, 53
    occipital, 49, 52
Cribriform plate, 48, 54
Cricothyroid, 69
Crista, 139
Crista galli, 48, 54
Crus of diaphragm, 121, 131,
    133, 134
Cuboidal epithelium, 1, 4–5, 38
Cuboid tarsal, 64
Cuneiform, 64
Cytoplasm, 41

D
Deep femoral artery, 120–121
Deep femoral vein, 120–121
Deep fibular nerve, 88
Deep peroneal nerve, 88, 153
Deltoid, 59, 66–67, 71, 74,
    78–79, 145
Deltoid tuberosity, 60
Dendrite, 18
Dense irregular connective
    tissue, 9
Dense regular connective
    tissue, 9
Dermis, 12
Descending aorta, 158
Developing bone, 15
Dialysis, 138
Diaphragm, 148
    in cat, 110, 112–114
    in fetal pig, 126–129
    in rat, 131, 133
Diaphysis, 139
Diastole, late, 157
Diffusion, 138
Digastric
    in cat, 94–96
    in human, 69, 143
Digital arteries, 88
Digits, bones of, 61, 64
Disks, intervertebral, 55
Dorsal horn, spinal cord, 20
Dorsalis pedis artery, 88
Dorsal ramus
    of lumbar nerve, 150
    of spinal nerve, 103
    of thoracic nerve, 109,
        127, 150
Dorsal root ganglion, 19
Dorsal venous arch, 159
Ductus arteriosus, 127, 160

Ductus deferens, 40
Ductus venosus, 127–129, 160

E
Ear, 140
Elastic artery, wall of, 8
Elastic cartilage, 11
Elastic tissue, 8
Elastin fiber, 8, 11
Endoplasmic reticulum, 136, 137
Endorphin, 161
Endothelium, 29
Eosinophil, 25, 156
Eosinophilic myelocyte, 15
Epicondyle, 92
    defined, 139
    femoral, 62
    humeral, 60
Epicranial aponeurosis, 68
Epicranius, 68, 142
Epidermis, 12
Epididymis, 40
    in cat, 119
    in fetal pig, 128
    in rat, 132
Epigastric artery, 130
Epiglottis, 70, 97, 114
Epiphyseal plate, 15
Epiphysis, 139
Epithelium, 1, 4–7, 22, 32, 36,
    38–41, 43
Epitrochlearis, 98, 107, 115
Erector spinae, 150
Erythroblast, 15
Erythrocyte, 27, 156
Esophagus, 32, 70
    in cat, 97, 112–114
    in fetal pig, 127
Ethmoid bone, 46, 48–50,
    52–54, 140
Eustachian tube, 114
Exocrine cells, pancreatic, 23
Exocytosis, 138
Extensor carpi digitorum
    lateralis, 100
Extensor carpi radialis, 98–99,
    101, 123
Extensor carpi radialis brevis,
    81, 100, 147
Extensor carpi radialis longus,
    80–81, 100, 147
Extensor carpi ulnaris
    in cat, 99, 100–101
    in fetal pig, 123
    in human, 81, 147
Extensor digiti minimi, 101
Extensor digitorum, 81, 147
Extensor digitorum brevis, 88
Extensor digitorum communis,
    100–101, 123
Extensor digitorum lateralis,
    101, 123
Extensor digitorum longus
    in cat, 104–105
    in human, 66, 85–86, 153
Extensor digitorum tendons, 101
Extensor expansion, 88

Extensor hallucis brevis, 88
Extensor hallucis longus,
    85, 88, 153
Extensor indicis, 101, 147
Extensor indicis proprius, 101
Extensor pollicis brevis,
    100–101, 147
Extensor pollicis longus, 147
Extensor retinacula
    in cat, 100–101, 105
    in human, 81, 85–86
External auditory meatus, 47,51, 91
External iliac artery, 118, 120–121,
    129, 158
External iliac vein, 118, 120,
    121, 159
External inguinal ring, 119
External intercostal, 74, 103, 148
External jugular vein, 160
    in cat, 94–95, 108–109,
        111, 114
    in fetal pig, 123, 125–126
External oblique
    in cat, 103, 106–107, 115
    in fetal pig, 122, 124
    in human, 66–67, 74–76, 148
    in rat, 130
External occipital crest, 49
External occipital protruberance,
    49, 91
Eye, 73, 141

F
Face
    bones of, 46, 140
    musculature of, 68, 142
Facet, defined, 139
Facial nerve, 94–96, 142, 154
Facial vein, 94–95
Facilitated diffusion, 138
False ribs, 58, 140
Fascia
    in cat, 98, 100–106
    in fetal pig, 122, 124
    in human, 66, 82–84
    in rat, 132
Femoral artery, 106, 118–121, 158
Femoral nerve, 106, 118–119,
    121, 152
Femoral triangle, 118–119
Femoral vein, 106, 118–121, 159
Femur
    in cat, 90–93
    in human, 62, 140
Fetal circulatory system, 160
Fetal pig, 122–129
Fibers, nerve, 19
Fibroblast, 8–9
Fibrocartilage, 10
Fibrocollagenous bundle, 41
Fibula
    in cat, 90, 93
    in human, 63, 140
Filtration, 138
First dorsal interosseous, 88
Fissure, defined, 139
Flagella, 136, 137

Flexor carpi radialis, 80, 98–99, 147
Flexor carpi ulnaris, 80–81, 98–100, 147
Flexor digitorum longus, 87, 106, 153
Flexor digitorum profundus, 99, 147
Flexor digitorum superficialis, 80, 147
Flexor hallicus longus, 63–64, 104–106, 124, 153
Flexor pollicis longus, 147
Flexor retinaculum, 87, 98–99
Floating ribs, 58
Foliate papillae, 97
Follicle stimulating hormone (FSH), 161
Fontanel, defined, 139
Foot
  arteries of, 88
  bones of, 64
  musculature of, 88
Foramen lacerum, 48–49
Foramen magnum, 47–49, 52
Foramen ovale, 48–50
Foramen rotundum, 48–50
Foramen spinosum, 48–50
Foramina
  defined, 139
  pelvic bones, 61, 62
  skull, 46–50
  spinal, 55–57
Fossa
  canine, 53
  cerebellar, 52
  cerebral, 52
  condylar, 49, 52
  coronoid, 60
  defined, 139
  glenoid, 59
  humeral, 60
  iliac, 61
  incisive, 53
  infraspinous, 59
  jugular, 48
  mandibular, 49, 51
  olecranon, 60
  pubic, 61
  scapular, 59
  subscapular, 59
  supraspinous, 59
Fovea, 139
Frenulum, 69
Frontal bone
  in cat, 91–92
  in human, 46–50, 140
Frontal border, 51
Frontal crest, 49
Frontalis, 66, 68
Frontal notch, 46
Frontal process, 47, 50, 53, 91
Frontal sinus, 48
Frontozygomatic suture, 47
FSH (Follicle stimulating hormone), 161
Furrows, parietal, 51

G
Gallbladder, 36
Ganglion cell, 19, 20
Gastric pits, 33
Gastrocnemius, 66
  in cat, 104–106
  in fetal pig, 124
  in human, 67, 84–87, 153
Gastroepiploic vein, 118, 159
Genioglossus, 69, 143
Geniohyoid, 69
Genitofemoral nerve, 121
Germinativum, 39
GH (Growth hormone), 161
Gladiolus, 59
Gland. *See Specific glands*
Glenoid fossa, 59
Glomerulus, 37–38
Glossopharyngeal nerve, 144, 154
Gluteal tuberosity, 62
Gluteus maximus
  in cat, 104–105
  in fetal pig, 122, 124
  in human, 67, 77, 83–84, 151
Gluteus medius
  in cat, 104–105
  in fetal pig, 122, 124
  in human, 67, 83–84, 151
Gluteus minimus, 83, 151
Goblet cell, 5, 33–34
Golgi apparatus, 136, 137
Gonadal vein, 159
Gracilis
  in cat, 106
  in human, 66–67, 82, 84, 87, 152
Granule, cytoplasmic, 25
Granulocyte, 156
Granulosa cells, 42
Greater omentum, 116
Greater palatine canal, 53
Greater sciatic notch, 62, 141
Greater trochanter, 62, 93
Greater tubercle, 60
Greater wing of sphenoid, 48, 49
Great saphenous nerve, 106
Great saphenous vein, 159
Growth hormone (GH), 161
Gynecoid pelvis, *versus* android, 141

H
Hair cell, 21
Hair follicle, 13
Hamate carpal, 61
Hamulus, 139
Hand
  bones and processes of, 61
  musculature of, 147
Hassall's corpuscle, 30
Haversian canal, 14
Haversian system (osteon), 14
Head. *See also* Face; Skull
  major veins of, 159–160
  muscles of cat, 94–97
  muscles of human, 68, 144

Heart
  cardiac cycle, 157
  cardiac muscle, 17, 65
  in cat, 110–112, 114
  in fetal pig, 126–129
  in rat, 131
Hemiazygos vein, 159
Hemorrhoidal artery, 129
Hepatic portal vein, 118, 159
Hepatic vein, 159
Hepatocyte, 35
Hip musculature, 83–84, 150–151
Horizontal plate, 52
Hormones, pituitary gland, 161
Human musculature. *See Musculature, human*
Human skeleton. *See Skeleton, human*
Humeral artery, 109
Humerus
  in cat, 90, 92
  in human, 60, 140
Hyaline cartilage, 10–11
Hyoglossus, 69
Hyoid bone, 69, 140
Hypogastric artery, 120
Hypogastric vein, 120
Hypoglossal canal, 49
Hypoglossal nerve, 143, 154
Hypoglossus, 143

I
Ileum, villi of, 33–34
Iliac artery
  in cat, 118, 120–121
  in fetal pig, 129
  in human, 158, 160
  in rat, 133
Iliac crest, 61–62
Iliac fossa, 61–62
Iliac spine, 61–62
Iliacus, 82, 151
Iliac vein
  in cat, 118, 120–121
  in human, 159
Iliocostalis
  in cat, 102–103
  in human, 72, 150
Iliolumbar artery
  in cat, 118, 120–121
  in rat, 133–134
Iliolumbar vein
  in cat, 118, 120
  in rat, 133–134
Iliopectineal line, 61–62
Iliopsoas, 66, 106, 150–151
Iliopubic eminence, 61–62
Iliotibial band, 83, 104
Ilium, 61, 92, 93
Incisive bone, 91
Incisive foramen, 49
Incisive fossa, 53
Incus, 48, 140
Inferior concha, 46
Inferior constrictor, 70
Inferior extensor retinaculum, 88

Inferior meatus, 53
Inferior nasal bone, 140
Inferior oblique, 73
Inferior orbital fissure, 49
Inferior pubic ramus, 61–62
Inferior rectus, 73
Inferior sagittal sinus vein, 159
Inferior vena cava, 113, 127, 160
Infraorbital foramen, 46, 50, 53
Infraspinatus, 67, 71, 78, 102, 145
Infraspinous fossa, 59, 92
Inguinal ring, 119
Innervation, 17
Innominate artery, 112, 127, 158
Innominate vein, 111, 114, 160
Intercalated disk, 17
Intercondylar fossa, 62
Intercostal artery, 131
Intercostal muscle, 74, 103, 148
Intercostal nerve, 148
Intercostal space, 139
Intercostal vein, 131, 159
Intermediate cuneiform, 64
Internal auditory meatus, 48, 51
Internal iliac artery, 120, 121, 129
Internal iliac vein, 120, 159
Internal intercostal, 103, 148
Internal jugular vein
  in cat, 108, 111, 114
  in fetal pig, 126
  in human, 160
Internal mammary artery, 113
Internal mammary vein, 126, 131
Internal oblique
  in cat, 115
  in fetal pig, 122, 124
  in human, 74–76, 148
  in rat, 130
Internal occipital crest, 52
Internal occipital protuberance, 52
Internal spermatic artery, 119
Interosseous border, 60, 63
Interosseous nerve, 101
Interphase, 2
Interstitial cell, 39
Intertrochanteric crest, 62
Intertubercular groove, 60
Intervertebral disk, 55
Intervertebral foramen, 55
Intestine. *See* Large intestine; Small intestine
Intrinsic muscles, human tongue, 143
Ischial ramus, 61–62
Ischial spine, 61–62, 141
Ischial tuberosity, 61–62, 93
Ischiocavernosus, 77, 149
Ischium
  in cat, 92–93
  in human, 61–62, 141
Islet of Langerhans, 23
Isometric ventricular contraction, 157
Isometric ventricular relaxation, 157
Isthmus of Fauces, 97

J

Jaw musculature, 142
Jugular foramen, 48–49
Jugular fossa, 48
Jugular notch, 52, 58–59
Jugular tubercle, 52
Jugular vein
    in cat, 94–97, 108
    in fetal pig, 123, 125–126
    in human, 160

K

Keratinized cell, 7
Kidney
    in cat, 116, 118, 119–121
    in fetal pig, 127–130
    in rat, 132–134
Kidney tubule, 4–5
Knee musculature, 84–87

L

Lacrimal bone
    in cat, 91
    in human, 46–47, 140
Lacrimal crest, 53
Lacrimal foramen, 46–47
Lacrimal fossa, 91
Lacunae, 10–11, 14
Lambda suture, 48
Lambdoidal margin, 52
Lambdoidal suture, 47
Lamina, vertebral, 56–57
Lamina propria, 29, 36, 38
Large intestine, 34, 116, 132
Larynx
    in cat, 95, 97, 114
    in fetal pig, 125–126
    in human, 70
    in rat, 131
Late diastole, 157
Lateral condyle, 62–63, 93
Lateral cuneiform, 64
Lateral epicondyle, 60, 92
Lateral malleolus, 88, 93
Lateral pterygoid, 50, 68, 142
Lateral pterygoid process, 49–50
Lateral rectus, 73
Latissimus dorsi
    in cat, 98–103, 107–109,
        111, 115
    in fetal pig, 122–123
    in human, 66–67, 71, 145
Leg
    musculature of, 82–87, 153
    veins of, 159
Lesser wing of sphenoid, 48–49
Leukocyte, 24–26, 156
Levator ani, 77, 149
Levator costa, 58, 148
Levator palpebrae superioris,
    73, 142
Levator scapulae, 69, 71, 78
Levator scapulae ventralis, 100
Levator veli palatini, 144
Leydig cell, 39
Ligament
    anterior cruciate, 62

costaclavicular, 59
    nuchal, 101–103
    patellar, 82, 85
    posterior cruciate, 62
Linea alba
    in cat, 107, 115
    in human, 66, 74, 76
Linea aspera, 62
Lingula, 53, 139
Liver, 35
    in cat, 116, 118–119
    in fetal pig, 128
    in rat, 130, 132
Longissimus
    in cat, 102, 103
    in human, 72, 144, 150
Long thoracic nerve, 108, 145
Long thoracic vein, 111
Longus capitis, 69
Loose connective tissue, 8
Lower extremities
    bones of, 62–64, 140
    major veins of, 159
    musculature of, 82–88,
        146–147, 151–153
Lumbar nerve, 119, 133, 155
Lumbar vein, 159
Lumbar vertebrae
    in cat, 90, 92–93
    in human, 55–56, 140
Lumbodorsal fascia
    in cat, 101–105
    in fetal pig, 124
Lumboscral plexus, 155
Lumen
    arterial wall, 28
    bladder, 39
    blood vessel, 7
    bronchiole, 32
    kidney tubule, 4, 5
    urethral, 39
Lunate carpal, 61
Lung
    in cat, 110–112
    in fetal pig, 126–129
    in rat, 131
Luteinizing hormone (LH), 161
Lymphatic vessel, 30
Lymph channel, 34
Lymph node
    in cat, 94–97, 111, 114
    in fetal pig, 123, 125
    histology, 29
Lymphocyte, 26, 156
Lymphoid follicle, 36
Lymph vessel, 117
Lysosome, 136–137

M

Macrophage, 4, 32
Malar bone. See
    Zygomatic bone
Malleolar fossa, 63
Malleolus, 93
Malleus, 48, 140
Mammary artery, 113
Mammary vein, 126, 131

Mandible
    in cat, 90–91, 96
    in human, 47, 53, 69–70, 140
Mandibular alveolus, 46, 47
Mandibular angle, 47
Mandibular condyle, 53
Mandibular condyloid process, 47
Mandibular coronoid process, 47
Mandibular foramen, 53
Mandibular fossa, 49, 51
Mandibular lingula, 53
Mandibular notch, 47
Mandibular ramus, 46, 53
Manubriosternal joint, 59
Manubrium, 58–59
Marrow cavity, 64
Masseter
    in cat, 94–96
    in fetal pig, 123, 125
    in human, 66, 68, 142
Mastication musculature, 68, 142
Mastoid margin, 52
Mastoid process, 47, 49, 51, 91
Matrix, hyaline cartilage, 10
Maxillae
    in cat, 90–91
    in human, 47, 49, 53, 140
Maxillary alveolus, 46, 47
Maxillary border, 50
Maxillary hiatus, 53
Maxillary process, 52
Maxillary sinus, 53
Meatus, 139
Medial condyle, 62–63, 93
Medial cubital vein, 111, 158
Medial cuneiform, 64
Medial epicondyle, 60, 92
Medial malleolus, 63, 88, 93
Medial pterygoid, 50, 68
Medial pterygoid process, 49–50
Medial rectus, 73
Median antebrachial vein, 158
Median crest, 57
Median nerve
    in cat, 96, 99, 101, 109,
        111–112, 114
    in human, 146–147
Median sacral artery
    in cat, 118, 120
    in fetal pig, 129
    in rat, 133
Mediastinal membrane, 128–129
Meissner's Corpuscle, 12
Membrane transport
    processsess, 138
Mental foramen, 46, 47, 53
Mental protuberance, 53
Mentum nuchae, 55
Mesenteric artery, 117,
    120, 121, 129
Mesenteric ganglion, 121
Mesenteric vein, 117, 159
Mesentery, 117
Metacarpal, 61, 90, 92, 140
Metaphase, 2
Metatarsal, 64, 90, 93, 140
Microfilament, 137

Microtubule, 2–3, 136–137
Microvilli, 136
Middle constrictor, 70
Middle ear, bones of, 140
Middle meatus, 53
Middle nasal concha, 54
Middle phalanx of digits, 61
Mitochondria, 136–137
Mitosis, phases of, 2–3
Monocyte, 26, 156
Monofid spinous process, 56
Motor endplate, 17, 135
Motor neuron, 19
Mucosa, 32, 34, 40
Multifidus, 102–103, 150
Muscle
    cardiac, 16
    smooth, 17
    striated, 16
Muscle fibers, 16
Muscularis (alimentary canal), 32
Muscularis (gallbladder), 36
Muscularis (large intestine), 34
Musculature, dissection specimen.
    See also Musculature, human
    cat, 94–109
    fetal pig, 122–129
    rat, 130–134
Musculature, human, 65–88
    anterior view, 66
    back and neck, 71–72
    chest and abdomen,
        74–76, 148
    eye, 73, 141
    head and neck, 68–70,
        142–144
    hip, 82–84, 150–152
    lower extremities, 82–88,
        146–147, 151–153
    pelvis, 77, 149
    posterior view, 67
    shoulder, 78–79, 145
    upper extremities,
        78–81, 145
Musculocutaneous nerve, 96, 99,
    109, 112, 114, 145–146
Myelinated nerve fiber, 20
Myelin sheath, 20
Myeloblast, 15
Myelocyte, 15
Mylohyoid
    in cat, 95–96
    in fetal pig, 125
    in human, 69, 143
Myoneural junction, 17

N

Nasal bone
    in cat, 91
    in human, 46, 47, 140
Nasal concha, 54
Nasal crest, 53
Nasal spine, anterior, 46, 53
Nasolacrimal crest, 53
Navicular carpal, 61
Navicular tarsal, 64
Navicular tuberosity, 64

Neck
    major veins of, 159–160
    musculature of, 69–72
Nerve ending, 13
Nerve fiber, 19–20
Nerves. *See Specific nerves*
Neuroglia, 18–19
Neurohypophysis, 23, 161
Neuron, 18–19
Neuronal process, 19
Neutrophil, 15, 24, 156
Neutrophilic stab cell, 15
Node of Ranvier, 19
Nuchal crest, 91
Nuchal ligament, 101–102
Nuchal line, 49
Nuclear envelope, 136
Nuclear membrane, 2
Nucleolus, 136–137
Nucleus, 136–137

O

Oblique capitis inferior, 144
Oblique muscle
    in cat, 103, 106–107, 113
    in fetal pig, 122, 124
    in human, 66, 74–75, 148
    in rat, 130
Obliquus inferior, 141
Obliquus superior, 141
Obturator foramen, 61–62
Obturator nerve, 106, 152
Occipital bone
    in cat, 91
    in human, 47–49, 52, 140
Occipital border, 51
Occipital condyle
    in cat, 91
    in human, 49
Occipitalis, 67–68
Occipital suture, 48
Occipitofrontalis, 142
Occipitosphenoid suture, 49
Occipitotemporal suture, 49
Oculomotor nerve, 141–142, 154
Odontoid process, 56–57
Olecranon fossa, 60
Olecranon process
    in cat, 92, 98
    in human, 60
Olfactory nerve, 154
Omohyoid, 69, 96, 143
Oocyte, 42
Optic canal, 73
Optic foramen, 46, 48–50
Optic nerve, 154
Orbicularis oculi, 66, 68, 142
Orbicularis oris, 66, 68, 142
Orbit, 50, 90, 91
Orbital border, 50
Orbital fissure, 46, 49, 50
Orbital plate
    of ethmoid bone, 54
    of frontal bone, 48–49
Orbital process, 52
Organ of Corti, 21
Os coxae, 140

Osmosis, 138
Osteoblast, 14
Osteocyte, 14
Osteon, 14
Ovarian artery, 118, 129, 134
Ovarian follicle, 42
Ovarian vein, 118, 134
Ovary
    in cat, 118
    in fetal pig, 129
    in human, 41
    in rat, 134
Ovum, 41–42
Oxytocin, 161

P

Pacinian Corpuscle, 13
Palate
    in cat, 97, 114
    musculature of, 144
Palatine bone, 49, 52, 140
Palatine canal, 53
Palatine foramen, 49
Palatine process, 53
Palatine tonsil, 31, 97
Palatoglossus, 69, 144
Palatopharyngeus, 144
Palmar arch vein, 158
Palmaris longus
    in cat, 98–99, 126
    in human, 80, 147
Pancreas, 23
    in cat, 116, 118–119
    in rat, 132
Papilla, 12–13, 97
Parathyroid gland, 22
Paraurethral gland, 39
Parietal bone
    in cat, 91–92
    in human, 46–48, 51, 140
Parietal cell, 33
Parietal notch, 51
Parotid gland, 37, 94, 111, 123
Patella, 63, 82, 85, 93, 140
Patellar ligament, 82, 85
Patellar tendon, 66
Pectineus, 66, 106, 152
Pectoantebrachialis, 95, 98,
    107, 115
Pectoralis major
    in cat, 98, 107, 115
    in fetal pig, 125
    in human, 66, 74, 145
Pectoralis minor
    in cat, 98, 107, 115
    in human, 74, 145
Pedicle, vertebral, 56, 57
Pelvis
    in cat, 90
    girdle and leg
        components, 140
    musculature of, 77, 149
    processes, 61–62
Penis
    in cat, 119, 121
    in fetal pig, 128
    in human, 77

    in rat, 130, 132–133
Pericardium, 110, 128
Peritoneum, 76, 115, 130
Peroneal nerve, 152–153
Peroneal retinacula, 86–87
Peroneal vein, 159
Peroneus brevis
    in cat, 104
    in human, 85–87, 153
Peroneus longus
    in cat, 104–105
    in human, 66–67,
        85–87, 153
Peroneus tertius, 85–86, 88, 153
Peroxisome, 137
Perpendicular plate, 54
Perpendicular plate of
    ethmoid, 46
Petrosal sinus vein, 159
Petrous portion of temporal
    bone, 48, 51
Phagocytosis, 138
Phalanges, 140
    in cat, 90, 92–93
    foot, 64
    hand, 61
Pharyngeal plexus, 144
Pharynx, 70, 144
Photoreceptors, 20
Phrenic nerve, 112–114, 148
Phrenicoabdominal artery, 121
Pig, fetal, 122–129
Pinocytosis, 138
Piriformis, 151
Pisiform carpal, 61
Pituitary gland, 23
    hormones of, 161
Placenta, 118, 160
Plantar arch vein, 159
Plantaris, 87, 153
Plasma membrane, 137
Platelet, 27, 156
Platysma, 68, 96
Podocyte, 37
Popliteal artery, 158
Popliteal surface, 62
Popliteal vein, 159
Popliteus, 153
Posterior cruciate ligament, 62–63
Posteriornasal spine, 49
Posterior pituitary
    (neurohypophysis), 161
Postural muscles, 150
Preerythroblasts, 15
Pregnant cat urogenital
    system, 118
Preperitoneal fat, 76
Primary follicle, 41
Primordial follicle, 41–42
Processes
    clavicle, 59
    lower extremities, 62–64
    pelvic, 61–62
    ribs, 58
    scapulare, 59–60
    skull, 46–54
    spinal, 55–57, 61

    sternum, 59
    upper extremities, 60–62
Prolactin (Prl), 161
Pronator quadratus, 80, 146
Pronator teres
    in cat, 98–99
    in human, 80, 146
Prophase, 2
Prostate gland, 41
    in cat, 119, 121
    in fetal pig, 128
Proximal extensor
    retinaculum, 105–106
Pseudostratified ciliated columnar
    epithelium, 5–6
Pseudostratified columnar
    epithelium, 40
Psoas, 151
    in cat, 120–121
    in rat, 133
Psoas major, 82
Pterygoid canal, 50
Pterygoid muscle, 68, 142
Pterygoid process, 49–50
Pubic arch, 61, 141
Pubic tubercle, 61
Pubis, 61–62, 93
Pudendal nerve, 149
Purkinje cells, 18
Pyramidal cell, 18
Pyramidal process, 52

Q

Quadratus lumborum, 150

R

Radial artery, 158
Radial fossa, 92
Radial nerve, 146–147
    in cat, 96, 99, 101, 108–109,
        111–112, 114
    in fetal pig, 127
Radial notch, 60
Radial process, 60
Radial tuberosity, 60
Radial vein, 158
Radius
    in cat, 90, 92
    in human, 60, 140
Ramus, 139
    mandibular, 46, 53
    pelvic, 61–62
Ranvier, Node of, 19
Rat, 130–134
Rectum, 77
    in cat, 117–121
    in fetal pig, 128–129
    in rat, 130, 132–134
Rectus abdominis
    in cat, 106–107, 115
    in human, 66, 74–76, 148
    in rat, 130, 132
Rectus femoris
    in cat, 106
    in human, 66, 82–83, 151
Red blood cell. *See* Erythrocyte
Red bone marrow, 15

Renal artery
     in cat, 118–121
     in fetal pig, 128
     in rat, 133–134
Renal corpuscle, 37
Renal vein, 159
     in cat, 120–121
     in fetal pig, 128–129
     in rat, 133–134
Reticular connective tissue, 8
Reticular fibers, 8
Retina, 20
Rhomboideus, 67
Rhomboideus capitis, 101, 123
Rhomboideus major, 71, 102, 145
Rhomboideus minor, 71,
     101–102, 145
Ribosome, 136–137
Ribs, 58, 140
Rods, 20
Root sheath, hair, 13
Rostrum, 50
Rotator cuff, 145

S

Sacral artery
     in cat, 118, 120
     in fetal pig, 129
     in rat, 133
Sacral canal, 57
Sacral foramen, 57
Sacral hiatus, 57
Sacral nerve, 155
Sacral vertebrae, 55, 57, 92
Sacrum, 55, 57, 61, 140, 141
Sagittal border, 51
Sagittal sinus
     grooves for, 52
     vein, 159
Sagittal suture, 47–48, 92
Salivary duct, 36–37, 123
Salivary gland, 36–37, 131
Saphenous vein, 106, 159
Sartorius, 66–67, 82–84, 87,
     104–106
Scalene, 148
Scalene tubercle, 58
Scalenus, 69, 108
Scalp, 13
Scaphoid carpal, 61
Scapulae
     in cat, 90, 92, 107
     in human, 59, 60, 78, 140
Scapular artery, 112–113
Scapular nerve, 145
Scapular vein, 111
Scarpa's fascia, 76
Sciatic nerve, 105
Sciatic notch, 62
Scrotum, 77
Sebaceous gland, 13
Secondary ovarian follicle, 42
Secretory vesicle, 136
Sella turcica, 48–49
Semimembranosus
     in cat, 104–106
     in fetal pig, 122, 124

in human, 67, 84, 87, 152
Seminal vesicle, 132–133
Seminiferous tubule, 39
Semispinalis, 150
Semispinalis capitis, 72, 144
Semitendinosus
     in cat, 104–106
     in fetal pig, 122, 124
     in human, 67, 84, 87, 152
Serous demilunes, 36
Serratus, 115, 122
Serratus anterior
     in cat, 115
     in fetal pig, 122
     in human, 66, 74, 145
Serratus anterior tuberosity, 58
Serratus ventralis, 108
Sertoli cell, 39
Sesamoid bone, 140
Shaft, 60
Shoulder
     bones of, 140
     musculature of, 71,
          78–79, 145
Sickle cell anemia, 27
Sigmoid colon, 129
Sigmoid sinus
     grooves for, 48, 51, 52
     vein, 160
Simple columnar epithelium, 5
Simple cuboidal epithelium, 1, 4–5
Simple squamous epithelium, 4
Sinus
     frontal, 48
     grooves for, 48, 52
Sinusoid, 35
Skeletal muscle, 16, 17, 89
Skeleton
     cat, 90–93
     human. See Skeleton, human
          terminology, 139
Skeleton, human, 46–64,
     140–141
          female versus male, 141
          lower extremities, 62–64
          pelvis, 61–62
          skull, 46–54
          spinal column, 55–57
          sternum and ribs, 58–59
          upper extremities, 60–61
Skin, 12
Skull
     bones of, 46–54, 140
     in cat, 91
     musculature of, 144
Small intestine, 33, 34
     in cat, 116–119
     in fetal pig, 128
     in rat, 132
Smooth muscle, 17, 32, 39
Soleal line, 63
Soleus, 66–67, 85–87, 153
     in cat, 104–106
     in fetal pig, 124
Somatotropin, 161
Spermatic artery
     in cat, 119, 121

in fetal pig, 128
     in rat, 133
Spermatic cord, 128
Spermatic vein, 119
Spermatocyte, 39
Spermatozoa, 39–40
Sphenoidal process, 52
Sphenoid bone, 46–50, 140
Sphenopalatine notch, 52
Sphincter ani externus, 149
Sphincter urethrae, 149
Spinal accessory nerves, 114,
     144–145, 154
Spinal column, 55–57
     components of, 140
     musculature of, 150
Spinal cord, 19, 20
Spinalis, 102–103, 150
Spinalis capitis, 72
Spinalis cervicis, 72
Spinalis thoracis, 72
Spinal nerve, 114, 155
Spindle fiber, 2, 3
Spine, 92, 139
Spinodeltoid
     in cat, 100–101
     in fetal pig, 122–123
Spinotrapezius
     in cat, 101–103
     in fetal pig, 122–123
Spinous process, 55–56, 91
Spiral line, 62
Spleen, 31
     in cat, 116, 118–119
     in fetal pig, 128
     in rat, 130, 132
Splenic vein, 159
Splenius capitis, 72, 101, 123
Spongy bone, 14, 64
Squamosal bone, 91
Squamosal border, 51
Squamosal suture, 47
Squamous epidermis, 12
Squamous epithelium, 4, 6–7
Stab cell, neutrophilic, 15
Stapes, 48, 140
Sternal angle, 58
Sternocleidomastoid, 66–69, 74,
     144, 148
Sternohyoid, 69, 143
     in cat, 94, 96
     in fetal pig, 125
     in rat, 131
Sternomastoid
     in cat, 94–96
     in fetal pig, 123, 125
Sternothyroid, 69, 94–95, 143
Sternum
     in cat, 90, 92, 95, 108
     in fetal pig, 125
     in human, 58–59, 140
     in rat, 130, 132–133
Stomach, 33
     in cat, 116, 118–119
     in fetal pig, 128
     in rat, 132
Stratified squamous epithelium, 6–7

Stratum basale, 12
Stratum corneum, 12
Stratum granulosum, 12
Stratum spinosum, 12
Striated muscle, 16, 89
Styloglossus, 69, 143
Stylohyoid, 69
Styloid process, 47, 49, 51,
     60, 69, 92
Stylomastoid foramen, 49, 91
Stylopharyngeus, 144
Subclavian artery, 158
     in cat, 112–113
     in fetal pig, 127
Subclavian groove, 58
Subclavian vein, 158
     in cat, 96, 99, 111, 114
     in fetal pig, 126
Sublingual salivary gland, 36
Submandibular salivary
     gland, 37, 94–95
Submaxillary gland
     in cat, 94–95, 111
     in fetal pig, 123, 125
Submucosa, 32, 34
Suboccipital nerve, 144
Subscapular artery, 113
Subscapular fossa, 59
Subscapularis, 79, 107–108,
     112, 145
Subscapular nerve, 109, 114, 145
Subscapular sinus, 31
Subscapular vein, 111, 114
Sulcus, 139
Superficial peroneal nerve, 88
Superficial transversus perinei, 77
Superior articular facet, 56–57
Superior articular process, 56–57
Superior constrictor, 70
Superior extensor retinaculum, 88
Superior nuchal line, 49
Superior oblique, 73
Superior orbital fissure, 49–50
Superior sagittal sinus vein, 159
Superior vena cava, 126–127, 160
Supinator, 80, 146
Suprameatal triangle, 51
Supraorbital foramen, 46, 50
Supraorbital notch, 50
Suprarenal vein, 159
Suprascapular nerve, 145
Suprascapular notch, 59
Supraspinatus
     in cat, 101–102
     in human, 71, 78
Supraspinous fossa, 59, 92
Surgical neck, defined, 139
Suture, cranial, 47–48
Symphysis pubis, 61–62, 77
Systole, atrial, 157

T

Talus, 64
Tarsal, 64, 90, 93, 140
Taste bud, 21
Tectorial membrane, 21
Telophase, 3

Temporal bone
in cat, 91
in human, 46–49, 51, 140
Temporal border, 50
Temporalis
in cat, 94, 96
in human, 67–68, 142
Temporal process, 46–47, 49–50
Temporozygomatic suture, 47
Tendon
Achilles, 64, 104–106, 124
calcaneal, 67, 86–87, 104
extensor digitorum, 101
foot, 88
patellar, 66
Tensor fascia latae
in cat, 104–106
in fetal pig, 122, 124
in human, 66, 82–83, 150
Tensor veli palatini, 107–108, 112, 144
Teres major
in cat, 107–108, 112
in human, 67, 71, 78, 145
Teres minor, 67, 71, 78, 145
Testis, 39
in cat, 119, 121
in fetal pig, 122, 124, 128
in rat, 130, 132
Thigh muscles, 82–84
Thigh musculature, 82–84, 151–152
Thoracic nerve
in cat, 96, 99
in fetal pig, 127
in human, 108, 145
Thoracic vein, 111, 159
Thoracic vertebrae
in cat, 90, 92
in human, 55–56, 58, 140
Thoracoacromial artery, 109, 112
Thoracoacromial nerve, 112
Thoracodorsal artery, 109
Thoracodorsal nerve, 145
in cat, 108–109, 111
in fetal pig, 127
Thoracodorsal vein, 111
Throat musculature, 143
Thumb, 61
Thymic corpuscles, 30
Thymus
in cat, 110–111
in fetal pig, 128
histology of, 30
in rat, 131
Thyrocervical artery, 113
Thyrohyoid, 69
Thyroid cartilage, 69
Thyroid gland, 69
in cat, 97, 114
in fetal pig, 126, 128
in rat, 131
thyroid follicle, 22
Thyroid gland follicle, 22
Thyroid stimulating hormone (TSH), 161

Tibia
in cat, 90, 93, 106
in human, 63, 85, 140
Tibial artery, 88, 158
Tibialis anterior
in cat, 104–106
in fetal pig, 124
in human, 66, 85–86, 153
Tibialis posterior, 63, 87, 106, 153
Tibial nerve, 152–153
Tibial tuberosity, 63, 93
Tibial vein, 159
Tongue, 69–70, 94, 97, 143
Tonsil, palatine, 31
Trachea
in cat, 95, 97, 111–114
columnar epithelium in, 6
in fetal pig, 125–128
in rat, 131
Transitional epithelium, 7, 38
Transport processess, membrane, 138
Transversalis abdominis, 115
Transversalis fascia, 76
Transverse foramen, 56–57, 91
Transverse jugular vein, 94–95
Transverse perinei, 149
Transverse process, 55–57, 91
Transverse scapular artery, 112–113, 127
Transverse scapular vein, 111
Transverse sinus
grooves for, 48, 52
vein, 160
Transversus abdominis, 74–75, 130, 148
Transversus thoracis, 148
Trapezium carpal, 61
Trapezius, 66–67, 69, 71, 74, 79, 144–145
Trapezoid carpal, 61
Triangle of ausculation, 145
Triangular carpal, 61
Triceps brachii
in cat, 99–102, 112
in fetal pig, 122–123
in human, 67, 78, 81, 146
Trigeminal nerve, 142–144, 154
Triquetrum carpal, 61
Trochanter, 139
in cat, 93
in human, 62
Trochlea, 60, 73, 92, 139
Trochlear nerve, 141, 154
Trochlear notch, 60
True ribs, 58, 140
Tubercle, 139
conoid, 59
intercondylar eminence, 63
jugular, 52
pubic, 61
rib, 58
talar, 64
Tuberosity, 60–62
defined, 139
of fifth metatarsal, 88
Tunica adventitia, 27–28

Tunica albuginea, 41
Tunica intima, 27–29
Tunica media, 27–29
Tunica muscularis, 39
Tunica propria, 39

U
Ulna
in cat, 90, 92
in human, 60, 140
Ulnar artery, 158
Ulnar nerve, 96, 98–99, 101, 109, 111–112, 114, 147
Ulnar notch, 60
Ulnar tuberosity, 60
Ulnar vein, 158
Umbilical artery, 119, 128–129, 160
Umbilical cord, 122
Umbilical vein, 128–129, 160
Umbilicus, muscles above and below, 76
Umbrella cell, 7, 39
Upper extremities
bones of, 60–61
major veins of, 158
musculature of, 78–81, 145–147
Ureter, 38
in cat, 118–121
in fetal pig, 128–129
in rat, 133–134
Urethra
in cat, 119, 121
in fetal pig, 128–129
histology, 39
in human, 77
Urethral oriface, 77
Urinary bladder, 39
in cat, 116–120
in fetal pig, 128–129
in rat, 132–134
Urogenital diaphragm, 77
Urogenital system
in cat, 118–119
in fetal pig, 128–129
in rat, 133–134
Uterine horn
in cat, 118
in fetal pig, 129
in rat, 134
Uterus
in cat, 117–118
in fetal pig, 129
in rat, 134

V
Vagina, 77
Vaginal oriface, 77
Vagus nerve, 144, 154
in cat, 111–114
in fetal pig, 126–127
Vas deferens, 119, 128
Vasopressin, 161
Vastus intermedius, 82, 151
Vastus lateralis
in cat, 105–106

in fetal pig, 122, 124
in human, 66–67, 82–84, 86, 151
Vastus medialis
in cat, 106
in human, 66, 82, 151
Vein. See also Specific veins
histology, 27
major and minor, 158–160
Vena cava
in cat, 111–114
in fetal pig, 126–127
Ventral canal, spinal cord, 20
Ventricular diastole, 157
Ventricular ejection, 157
Ventricular systole, 157
Vermiform appendix, 36
Vertebrae
in cat, 90–93
in fetal pig, 127
in human, 55–57, 150
Vertebral arches, 56
Vertebral artery, 112–113, 127
Vertebral column, musculature of, 150
Vertebral foramen, 56
Vertebral ribs, 58
Vertebral spinous process, 92
Vertebral vein, 160
Vertical plate, palatine bone, 52
Vibrissal barrels, 94
Villi of Ileum, 33, 34
Vomer bone, 46, 49, 140
Vomeronasal nerve, 154

W
White blood cell. See Leukocyte
White matter, 20
Willis, Circle of, 158
Wormian bones, 47, 140
Wrist musculature, 147

X
Xiphihumeralis, 107, 115
Xiphisternal joint, 59
Xiphoid process, 59. See also Sternum

Z
Zona fasciculata, 24
Zona glomerulosa, 24
Zona pellucida, 42
Zona reticularis, 24
Zygomatic arch, 47, 49, 90–91
Zygomatic bone
in cat, 91
in human, 46–47, 49, 50, 140
Zygomatic process, 50
of frontal bone, 91
of maxilla, 46, 53
of temporal bone, 47, 49, 51, 91
Zygomaticus, 66, 68
Zygomaticus major, 142
Zygomaticus minor, 142
Zymogen granules, 37